T0330100

Accessibility and Spatial Interaction

NECTAR SERIES ON TRANSPORTATION AND COMMUNICATIONS NETWORKS RESEARCH

Series Editor: Aura Reggiani, *Professor of Economic Policy, University of Bologna, Italy*

NECTAR (Network on European Communications and Transport Activities Research) is an international scientific, interdisciplinary association with a network culture. Its primary objective is to foster research collaboration and the exchange of information between experts in the fields of transport, communication and mobility.

NECTAR members study the behaviour of individuals, groups and governments within a spatial framework. They bring a wide variety of perspectives to analyse the challenges facing transport and communication, and the impact these challenges have on society at all levels of spatial aggregation.

This series acts as a companion to, and an expansion of, activities of NECTAR. The volumes in the series are broad in their scope with the intention of disseminating some of the work of the association. The contributions come from all parts of the world and the range of topics covered is extensive, reflecting the breadth and continuously changing nature of issues that confront researchers and practitioners involved in spatial and transport analysis.

Titles in the series include:

Transportation and Economic Development Challenges
Edited by Kenneth Button and Aura Reggiani

City Distribution and Urban Freight Transport
Multiple Perspectives
Edited by Cathy Macharis and Sandra Melo

Accessibility Analysis and Transport Planning
Challenges for Europe and North America
Edited by Karst T. Geurs, Kevin J. Krizek and Aura Reggiani

Smart Transport Networks
Market Structure, Sustainability and Decision Making
Edited by Thomas Vanoutrive and Ann Verhetsel

Accessibility and Spatial Interaction
Edited by Ana Condeço-Melhorado, Aura Reggiani and Javier Gutiérrez

Accessibility and Spatial Interaction

Edited by

Ana Condeço-Melhorado

Researcher Fellow, European Commission Joint Research Centre (JRC) Institute for Prospective Technological Studies (IPTS), Spain

Aura Reggiani

Professor, University of Bologna, Italy

Javier Gutiérrez

Professor, University Complutense of Madrid, Spain

NECTAR SERIES ON TRANSPORTATION AND COMMUNICATIONS NETWORKS RESEARCH

Edward Elgar

Cheltenham, UK • Northampton, MA, USA

Published by
Edward Elgar Publishing Limited
The Lypiatts
15 Lansdown Road
Cheltenham
Glos GL50 2JA
UK

Edward Elgar Publishing, Inc.
William Pratt House
9 Dewey Court
Northampton
Massachusetts 01060
USA

A catalogue record for this book
is available from the British Library

Library of Congress Control Number: 2014950735

This book is available electronically in the ElgarOnline.com
Economics Subject Collection, E-ISBN 978 1 78254 073 1

ISBN 978 1 78254 072 4

Typeset by Servis Filmsetting Ltd, Stockport, Cheshire
Printed and bound in Great Britain by T.J. International Ltd, Padstow

Contents

Contributors

Pelayo Arbués, University of Oviedo, Spain.

José Baños, University of Oviedo, Spain.

Simone Caschili, University College London, UK.

Ana Condeço-Melhorado, European Commission Joint Research Centre (JRC), Institute for Prospective Technological Studies (IPTS), Spain.

Andrea De Montis, University of Sassari, Italy.

Giacomo Galiazzo, University of Bologna, Italy.

Urban Gråsjö, University West, Sweden.

Javier Gutiérrez, University Complutense of Madrid, Spain.

Kingsley Haynes, George Manson University, USA.

Adelheid Holl, CSIC – Spanish National Research Council, Spain.

Charlie Karlsson, Jönköping International Business School, Sweden.

Rajendra Kulkarni, George Manson University, USA.

Matías Mayor, University of Oviedo, Spain.

David Philip McArthur, University of Oslo, Norway.

Kai Nagel, Technical University of Berlin, Germany.

Thomas W. Nicolai, Technical University of Berlin, Germany.

John Östh, Uppsala University, Sweden.

Aura Reggiani, University of Bologna, Italy.

Paula Remoaldo, Minho University, Portugal.

Vitor Ribeiro, Minho University, Portugal.

María Henar Salas-Olmedo, University Complutense of Madrid, Spain.

Laurie A. Schintler, George Manson University, USA.

Roger Stough, George Manson University, USA.

Inge Thorsen, Stord/Haugesund University College, Norway.

Daniele Trogu, University of Cagliari, Italy.

Jan Ubøe, NHH Norwegian School of Economics, Norway.

Preface

NECTAR (Network on European Communications and Transport Activities Research) is a network of multidisciplinary researchers that brings together a wide range of perspectives on transportation, mobility and communication issues and their impact on society.

NECTAR organizes numerous thematic seminars and workshops together with a major conference every two years. Most of the chapters included here are a revised contribution of papers presented in a special session on 'Accessibility and Spatial Interaction' organized by the editors of this book under the 9th World Congress of Regional Science Association International held in June 2012 in Timisoara, Romania. Other internationally renowned authors in the field of accessibility research were also invited to contribute chapters to this book.

The editors would like to thank the contributors for their time and effort in preparing the chapters included here and their attentive response to the peer reviewer's comments. We are also grateful to Alan Sturmer (Executive Editor, Edward Elgar) for his support in organizing this book.

Ana Condeço-Melhorado, Aura Reggiani and Javier Gutiérrez
April 2014

1. Accessibility and spatial interaction: an introduction

Ana Condeço-Melhorado, Aura Reggiani and Javier Gutiérrez

1.1 THE CONTEXT: SPATIAL INTERACTION AND ACCESSIBILITY ANALYSIS

Spatial interaction refers to economic or demographic flows between different locations, generally through a transport infrastructure. It implies a complementarity between two places engaged in a supply–demand relationship which is subject to certain costs, such as cost of transport or transaction costs.

Spatial interaction models (SIMs), stemming from their analogy with Newton's law of gravity, are especially suited to study the spatial interaction between pairs of locations, where space is measured in the form of cost–utility function (at the aggregate or disaggregate level).

SIMs have a long history and have been used in a wide variety of contexts. Wilson (1970) gave SIMs theoretical strength by deriving them using the entropy maximization approach. Nijkamp (1975) offered an economic framework by deriving SIMs from a cost-minimization approach and, finally, Anas (1983) demonstrated the formal equivalence between SIMs and logit models and microeconomic theory. All in all, SIMs in all their forms (unconstrained, singly constrained and doubly constrained) can be interpreted in macroeconomic and microeconomic terms, where a key behavioural element is provided by the cost function parameter (for a review, see Reggiani, 2014).

The accessibility function can be directly derived from SIMs and it therefore contains SIM behavioural cost components. Its analytical formulation is consistent with the accessibility concept and equation provided by Hansen (1959), the first author to apply a gravity-type model to the study of accessibility, defining it as 'potential of opportunity for interaction' (p. 73). In other words, accessibility of a given place is related to the importance of the opportunities available for interaction, as well as with the distance needed to reach those opportunities. Since then, accessibility analyses have addressed the issue of spatial interaction, while increasing the level of complexity: (1) by including other factors in addition to distance or the weight of places (van Wee et al., 2001); and (2) by using more detailed data and calibrating the models in a way that correctly represents the deterrence of distance, that is, the effort needed to overcome the distance between locations (Baradaran and Ramjerdi, 2001; Martínez and Viegas, 2013; Reggiani et al., 2011). Sensitiveness to cost–distance factor is known in accessibility studies as distance decay and it differs with factors such as individual perceptions, type of transport infrastructure or spatial characteristics of places under analysis. Doubly constrained or unconstrained SIMs are commonly used to estimate different types of distance decay.

There is no doubt that spatial interaction is related to accessibility, since improved accessibility will increase the spatial interaction between places. However this relation is often asymmetric, since some places take more advantage than others of better accessibility. Accessibility improvements can emerge from transport infrastructure developments or from land use changes, that is, increased number of jobs in a certain location. Accessibility indicators are able to capture changes in both components (transport infrastructure and land use) and for this reason their performance has been recognized in the framework of sustainability policy, in light of best practice planning and decision-making processes where evaluation methods, such as cost–benefit analysis, multi-criteria analysis and risk analysis, can embed accessibility results.

Given their spatial nature, accessibility indicators are very sensitive to the spatial unit of analysis (what is known as the modifiable area unit problem, MAUP). The MAUP will influence the goodness of fit of the calibration process (with spatial interaction data) and determine the role of self-potential. 'Self-potential' refers to the intra-zonal level of spatial interaction, which can be significant and even outweigh interaction between zones, especially in the most urbanized locations (Frost and Spence, 1995). The difficulty of estimating self-potentials or self-interactions in accessibility studies due to lack of data on transport networks within zones is widely acknowledged. This issue, which particularly affects gravity-type accessibility indicators, can be minimized by using smaller zones and new techniques to improve computational performance. Nowadays new sources of data are available that better represent intra-zonal and inter-zonal spatial interactions. This is true of data from location-sharing devices that report information with high spatial and temporal detail. The use of this data is a promising way of improving our understanding of mobility patterns and paves the way towards a stronger theoretical analysis of accessibility indicators, with reference to their dynamic framework.

Accessibility varies in space, since it is related to economic activity variables such as population, income, employment, and so on, and to the distance or cost and time of reaching these activities. In short, accessibility can be interpreted as the spatial arrangement of economic opportunities. Accessibility can be high around metropolitan areas and decreases as we move away from urban agglomerations. Accessibility also decreases whenever there is a barrier, as in the case of national borders. Several studies have found that national borders exert a negative influence on issues such as trade flows. Another source of variation in accessibility patterns is related to social groups: for example, young people are able to move faster and more easily than elderly people. Aspects such as the barrier effect (such as national borders) and the mobility patterns of different social groups can be included in accessibility analysis as methodological improvements.

Accessibility is associated with higher productivity levels, as regions with high accessibility benefit from lower transport costs and agglomeration economies. Accessibility also concerns patent output, regional economic growth, new firm formation, labour and export performance of regions.

It should be noted that, given the increasing relevance of accessibility issues, mostly in this globalized space-economy, a great number of review studies have been published that also show the potential and role of this tool for best practice and planning (see, among others, in the last decade: Becker et al., 2008; Dentinho et al., forthcoming; De Montis and Reggiani, 2012, 2013; Geurs and van Wee, 2004; Geurs et al., 2012; Gutiérrez et al., 2010; Levinson and Krizek, 2008). In this context, we aim to go beyond these accessibility analysis studies and to reflect on the specific issue of the link between spatial interaction and accessibility from the theoretical, methodological, empirical and policy analysis perspectives.

On the basis of the above, the rationale of this book is to present a collection of recent studies modelling and discussing spatial interaction by means of accessibility indicators. Table 1.1 summarizes the main characteristics of all the studies included in this book, focusing particularly on the applied measurement of accessibility.

The chapters have been organized in three parts: Part I deals with methods and data sources used to estimate spatial interaction through accessibility indicators. Part II is devoted to the social and spatial dimension of interaction. Finally, Part III of this book is dedicated to accessibility as a driver of spatial interaction, exploring the relationship between accessibility and regional economic performance.

Table 1.1 Overview of chapters included in this book

Chapter	Object	Accessibility measure	Attractiveness	Variable in the decay function	Form of the decay function	Spatial interaction data	Spatial units	Spatial context
J. Östh, A. Reggiani, G. Galiazzo (Ch. 2)	Estimation of distance decay parameters	Potential accessibility	Jobs	Cartesian distances	Power and exponential decay	Commuting flows	Municipalities	Sweden
D.P. McArthur, I. Thorsen, J. Ubøe (Ch. 3)	Impacts of accessibility to labour markets, measured in terms of distributions of employment and population	Weighted-average distance	Number of jobs, adjusted for the competition for jobs	Generalized transport costs	Logistic function	Hypothetical data on migration flows	Zones	Simulation
T.W. Nicolai, K. Nagel (Ch. 4)	Influence of spatial resolution, congestion and transport mode on accessibility outcomes	Logsum indicator	–	Travel time	Exponential decay	Home–work–home commuting plans	Origins: cells of different sizes and zones centroids; Destinations: network nodes	City of Zurich

L.A. Schintler, R. Kulkarni, K. Haynes, R. Stough (Ch. 5)	Spatial and temporal variations in the mobility patterns of individuals	Indicators based on the bipartite network modelling and mobility sheds	Counties that share common visitors	–	–	Brightkite location-sharing services data	Counties	Lower 48 states in the US
A. De Montis, S. Caschili, D. Trogu (Ch. 6)	Spatial dependence of accessibility in US counties	Potential accessibility	Number of commuters	Travel cost	Exponential and power decay	Commuting flows	Counties	US
M.H. Salas-Olmedo, A. Condeço-Melhorado, J. Gutiérrez (Ch. 7)	Border effects in the European Union	Potential accessibility	Gross domestic product	Travel time	Power decay	Trade flows	Countries	EU
V. Ribeiro, P. Remoaldo, J. Gutiérrez (Ch. 8)	Access of elderly people to bus stops	Accumulated opportunities	Bus stations	Travel time	–	Observed travel patterns of elderly people	Parishes	City of Braga (Portugal)

Table 1.1 (continued)

Chapter	Object	Accessibility measure	Attractiveness	Variable in the decay function	Form of the decay function	Spatial interaction data	Spatial units	Spatial context
P. Arbués, M. Mayor, J. Baños (Ch. 9)	Productivity impacts of road infrastructure	Potential accessibility	Population	Mean travel distance	–	–	Provinces	Spain
A. Holl (Ch. 10)	Location characteristics and corporate productivity	Potential accessibility	Population	Travel time	–	–	Municipalities	Spain
U. Gråsjö, C. Karlsson (Ch. 11)	Review chapter	Potential accessibility	Innovation (e.g. research and development, patents), population, gross regional product, etc	Proximity	–	Various	Municipalities, functional regions	Sweden

1.2 ADVANCES IN MODELLING ACCESSIBILITY AND SPATIAL INTERACTION

Part I presents novel methodologies used to calibrate accessibility with spatial interaction data, as well as advances regarding the dynamic interaction between transport and land use. It also reflects on the role of the unit of analysis in estimating accessibility and finally presents an analysis of mobility and accessibility patterns based on new sources of data, such as those from Global Positioning System (GPS)-equipped devices.

In Chapter 2, John Östh, Aura Reggiani and Giacomo Galiazzo highlight the importance of correctly estimating distance decay in accessibility analysis. The authors explore two commonly used distance decay functions – the exponential and the power function – and compare three different methods to estimate the distance decay parameter. In particular, the authors adopt two SIMs – the unconstrained and the doubly constrained SIMs – to calibrate spatial commuting flows between Swedish municipalities. The third method used to estimate the distance decay parameter is the half-life model embedded in SIMs, which is a new approach in such estimation studies. The main advantage of this method is the reduced amount of data required to estimate the distance decay parameter. The authors also reflect on how distances between places are specified, testing mean and median distance values as a way of determining the sensitiveness of distance decay methods to the size of geographical units (MAUP). Based on different distance decay estimates, the authors calculate several accessibility indicators for the spatial context of Sweden over a 15-year period, concluding that the accessibility model least sensitive to the MAUP is that which is calculated with the power function. The half-life model can be used as an alternative to the exponential function when flow data is missing, since similar accessibility patterns emerged from these two models.

Changes in transport infrastructure will reduce distance-related costs and this will have a positive impact on interaction between places. On the other hand, a change in the size of places motivated by an increase in the number of jobs, for example, will also increase interaction by attracting and generating new flows of people, goods and immaterial flows to that place due to the development of new residential areas and (re)location of firms. In most cases, a change in one factor cannot be dissociated from the others; thus, a change in infrastructure will lead to a change in size. Chapter 3 of this book deals with the complex interaction that could arise when transport networks are expanded or improved. David Philip McArthur, Inge Thorson and Jan Ubøe use a spatial general equilibrium model to estimate the impact of changing accessibility to a series of places as a result of a new bridge. Impacts are measured in terms of population, employment and migrations (of people and relocation of firms), with particular focus on urban and rural dimensions. They show that although accessibility may be improved in the short run due to reduced transport costs, in the long run population and jobs located in rural areas may fall. Some rural zones close to the central city may also register a decrease in terms of employment and an increase in population, reflecting a change from a rural to a suburban zone.

In Chapter 4, Thomas Nicolai and Kai Nagel analyse the influence of spatial disaggregation in accessibility computations, as well as the effect of considering congestion and different transport modes over accessibility levels. The authors use a logsum indicator to measure location accessibility, which accounts for the number and distribution of opportunities and the transport infrastructure. They use the city of Zurich as a case study, and couple a land use model (UrbanSim) with a transport model (MATSim), to perform a traffic simulation to estimate congested travel times for different transport modes (car, bicycle and walking). Regarding the spatial resolution of origin locations, the authors compare two approaches: zone-based and cell-based. For the zone-based approach, the centroid represents the origin location, while for the cell-based approach the location is represented by points that in turn depend on the size of the cell. Opportunities, meanwhile, are aggregated on the nearest node on the road network. The effect of spatial resolution is evaluated in terms of quality of the results and computational performance. The authors recommend the use of the cell-based approach, since it removes the controversial assumption, taken by the zone-based approach, of homogeneous accessibility within zones. Furthermore the cell-based approach showed similar computational efforts compared to the zone-based approach and yielded more meaningful results. However there is a disutility in improving spatial resolution in the cell-based approach, since at certain levels of high resolution the computational performance does not compensate in terms of improved quality of results. Apart from spatial resolution considerations, the authors conclude that the effect of congestion clearly changes accessibility levels in Zurich. In terms of transport modes, they reach the interesting conclusion that during peak hours in the urban core, accessibility by bicycle is similar to accessibility by car. In contrast, accessibility when walking is clearly worse than the other modes.

Laurie A. Schintler, Rajendra Kulkarni, Kingsley Haynes and Roger Stough (Chapter 5) provide a novel contribution on how new sources of data can be used to analyse mobility and accessibility patterns. Specifically, they use data from location service devices in the USA between 2008 and 2010. They present a set of indicators that are able to characterize the degree of mobility of US counties, the strength of ties between counties and the centrality of counties within the network of locations. Their results showed that central counties also have a higher number of individuals using these location service devices and that densely populated regions tend to be highly mobile. Their study also looks at travel distance patterns (national or regional) of individuals at a certain location, showing that tourist destinations generally present a larger number of domestic trips than regional trips. Finally, they identify distinct mobility sheds in the US, defined as regions in which the same set of people are travelling, concluding that these communities are usually in close proximity in spatial terms.

1.3 THE SOCIAL AND SPATIAL DIMENSION OF ACCESSIBILITY

Part II comprises three chapters on spatial patterns on accessibility and interaction from different perspectives. In Chapter 6, Andrea De Montis, Simone Caschili and Daniele Trogu analyse the role of spatial dependence in the accessibility of US counties, considering the extent to which the accessibility of a county depends on the characteristics of neighbouring counties. Using autocorrelation analysis the authors address two fundamental questions: (1) whether counties with high accessibility positively influence adjacent counties; and (2) the spatial association between accessibility levels and other socio-economic variables (population and income per capita). To answer the first question, the authors perform a univariate spatial autocorrelation analysis with the Moran I and the Local Indicator of Spatial Autocorrelation (LISA). Results show a strong spatial dependence on the accessibility of US counties, meaning that counties with similar levels of accessibility tend to cluster together. Furthermore they show the distribution of clusters of high and low accessibility levels across the country. To answer the second question the authors used a bivariate autocorrelation analysis, concluding that high accessibility clusters are located in large metropolitan regions, while low accessibility clusters are located in central areas of the US characterized by lower population density. Similar patterns were found for the association between accessibility and income, but in this case the authors also found clusters with negative spatial autocorrelation. This occurs in regions with higher income and less accessibility than surrounding counties, or in regions with lower income and higher accessibility than adjacent pairs.

María Henar Salas-Olmedo, Ana Condeço-Melhorado and Javier Gutiérrez (Chapter 7) reflect on the decrease of spatial interaction resulting from national borders. This is known as the border effect and has been measured in several studies using spatial interaction models and focusing on trade flows between countries. However the literature is unclear with regard to the magnitude of the border effect and it is argued that part of this controversy is due to how distances have been measured by previous authors. Following this rationale, the chapter makes two main contributions to the existing literature: (1) understanding the role of different distance metrics when measuring border effects; and (2) including border effects in the accessibility computation. Border effects are analysed for a set of European countries considering different distance metrics such as Euclidean distance, network distance, travel time and generalized transport costs. They found a higher preference in all countries for national rather than international trade. This is also true for all distance metrics; however, border effects increase with the complexity of the distance metric (higher when travel times and generalized transport costs are considered). Border effects are generally higher in peripheral countries and may reflect less competitiveness to trade in international markets. Finally, the authors introduce the border effect in the market potential indicator in order to capture not only the decrease of distance-related interaction, but also the effect of national borders on spatial interaction.

In Chapter 8, Vitor Ribeiro, Paula Remoaldo and Javier Gutiérrez analyse the accessibility of elderly people to public transport. They argue that urban characteristics such as street slope angle and the speed at which the elderly people walk should be included when modelling accessibility of the elderly, since this can greatly affect their ability to reach public transport. The authors analyse the accessibility of the elderly to bus stops in the municipality of Braga (Portugal) and compare different measurements, including slope angle and walking speed, concluding that the inclusion of street slope angle and walking speed provides better estimates of the accessibility to bus stops for this group. Based on these results, they argue that these factors should be considered in studies dealing with transport-related social exclusion and in the analysis of spatial inequalities.

1.4 ACCESSIBILITY AS A DRIVER OF SPATIAL INTERACTION

The first two chapters of Part III look at the relationship between accessibility and productivity. Pelayo Arbués, Matías Mayor and José Baños (Chapter 9) focus on measuring the output effect of road transportation infrastructure in Spanish provinces between 1997 and 2006. They estimate a production function using a panel of data from Spanish provinces in order to account for marginal productivity effects within a province and to document the existence of spillover effects outside the provincial boundaries through the use of spatial econometrics methodologies. Road infrastructure endowment indicators are measured in three different ways: using an accessibility measure, the traditional road stock indicators and a variable that attempts to adjust the stock indicator to the degree of utilization. The authors found strong evidence of the positive impact of better road accessibility on the domestic economy of a province. The results also showed that improved accessibility would increase the productivity of neighbouring areas even more than the spatial unit.

In Chapter 10, Adelheid Holl presents an empirical analysis of the relationship between location characteristics and productivity using data from Spanish firms. Both urban agglomeration and better transport accessibility can provide firms with access to denser and larger markets and this may help improve efficiency and increase productivity. The analysis distinguishes different location characteristics such as local population size, local population density and accessibility. The results show a significant positive productivity effect for all three location characteristics, particularly accessibility, and the author concludes that transport accessibility better captures the benefits of location.

The concluding chapter is a contribution from Urban Gråsjö and Charlie Karlsson and analyses the potential of accessibility in spatial economics. They look at different empirical studies conducted in a Swedish context where accessibility measures were used to explain patent output, regional economic growth, new firm formation and firm dynamics, the emergence of new export products and labour mobility. All studies reviewed show the positive impact of accessibility on market size and knowledge sources. Accessibility is considered a useful tool for spatial economic studies because: (1) it is related to economic theories such as spatial interaction and random choice; (2) it captures the spatial dependencies between locations and thus spillover effects; and (3) it gives a better representation of space-economy evolution by providing useful insights for policy-making.

This book aims to address a wide variety of issues and perspectives linked to the accessibility concept in order to outline its analytical fruitfulness – due to its transversal nature – as well as its performance in different empirical applications in light of specific policy strategies. Nevertheless, the potential of accessibility analysis has still not been fully exploited. In the near future new topics will certainly be addressed with novel approaches using accessibility tools. Certainly the impact of new sources of information, such as information and communication technologies (ICT), will be as significant for accessibility studies as geographic information systems (GIS) were several years ago in terms of offering advanced findings, mainly with reference to the complex dynamic interaction between accessibility and space-economy.

REFERENCES

Anas, A. (1983), 'Discrete choice theory, information theory and the multinomial logit and the gravity models', *Transportation Research B*, **17** (1), 13–23.

Baradaran, S. and F. Ramjerdi (2001), 'Performance of accessibility measures in Europe', *Journal of Transportation and Statistics*, **4**, 31–48.

Becker, U., J. Böhmer and R. Gerike (eds) (2008), 'How to define and measure access and need satisfaction in transport', *Dresden Institute for Transportation and Environment (DIVU)*, 7, Dresden: University of Dresden.

De Montis, A. and A. Reggiani (eds) (2012), Special Issue on 'Accessibility and socio-economic activities: methodological and empirical aspects', *Journal of Transport Geography*, **25**.

De Montis, A. and A. Reggiani (eds) (2013), Special Issue on 'Analysis and planning of urban settlements: the role of accessibility', *Cities*, **30**.

Dentinho, T., K.T. Geurs and R. Patuelli (eds) (forthcoming), *Accessibility and Amenities*, Cheltenham, UK and Northampton, MA, USA: Edward Elgar.

Frost, M.E. and N.A. Spence (1995), 'The rediscovery of accessibility and economic potential: the critical issue of self-potential', *Environment and Planning A*, **27**, 1833–1848.

Geurs, K.T. and B. van Wee (2004), 'Accessibility evaluation of land-use and transport strategies: review and research directions', *Journal of Transport Geography*, **2**, 127–140.

Gutiérrez, J., A. Condeço-Melhorado and J.C. Martín (2010), 'Using accessibility indicators and GIS to assess spatial spillovers of transport infrastructure investment', *Journal of Transport Geography*, **18**, 141–152.

Hansen, W.G. (1959), 'How accessibility shapes land-use', *Journal of American Institute of Planners*, **25**, 73–76.

Levinson, D.M. and K.J. Krizek (2008), *Planning for Place and Plexus*, New York: Routledge.

Martínez, L.M. and J.M. Viegas (2013), 'A new approach to modelling distance-decay functions for accessibility assessment in transport studies', *Journal of Transport Geography*, **26**, 87–96.

Nijkamp, P. (1975), 'Reflections on gravity and entropy models', *Regional Science and Urban Economics*, **5**, 203–225.

Reggiani, A. (2014), 'Complexity and spatial networks', in M. Fischer and P. Nijkamp (eds), *Handbook of Regional Science*, Berlin and Heidelberg, Germany and New York, USA: Springer, pp. 811–832.

Reggiani, A., P. Bucci and G. Russo (2011), 'Accessibility and network structures in the German commuting', *Networks and Spatial Economics*, **11** (4), 621–641.

Van Wee, B., M. Hagoort and J.A. Annema (2001), 'Accessibility measures with competition', *Journal of Transport Geography*, **9**, 199–208.

Wilson, A.G. (1970), *Entropy in Urban and Regional Modelling*, London: Pion.

PART I

Advances in modelling accessibility and spatial interaction

2. Novel methods for the estimation of cost–distance decay in potential accessibility models

John Östh, Aura Reggiani and Giacomo Galiazzo

2.1 INTRODUCTION

For commuters, homebuyers and planners alike, accessibility, be it to jobs, services or recreation, is a highly sought-after but elusive amenity. Elusive not only because what is accessible for some may be inaccessible or unattractive to others, but also because accessibility can be measured differently, being dependent on a variety of calibration and specification decisions, and on issues relating to the quality and level of data aggregation.

Over the years various categories of accessibility measures have been developed, capturing the varying needs for and specifications of access in different fields. A commonly used classification of accessibility measures organizes the different measures into three types: cumulative opportunities measures, gravity-based or potential measures, and utility-based measures (Handy and Niemeier, 1997). Cumulative measures count the total amount of opportunities reachable within a predefined radius, cost or time. Potential measures weight opportunities by impedance, which is generally a function of travel time, distance or the cost of moving between locations. Utility-based measures consider accessibility to be the individual choice that creates the greatest utility for each individual. In spatial science, potential measures of accessibility are more common than the other measures.

One of the general limitations to accessibility measures is that aggregate level data rather than disaggregate level data is being used in the analysis (Kwan, 1998). This is partly because aggregate level data is easier to obtain, but also because trip-distribution models are difficult to use on highly disaggregated units. A problem with using aggregate level data is that it forces researchers to assume that individual and workplace behaviour is uniform within the aggregate area, and that jobs and workers are distributed uniformly within each aggregated unit (Östh, 2011). In reality this is rarely the case. In most cases the spatial arrangement of areas distorts the underlying behaviour of what is being measured to an unknown extent – this is a problem known as MAUP (modifiable areal unit problem; see Openshaw, 1984; Fotheringham and Wong, 1991). This means that bias is introduced not only when choosing how to measure interaction within and between areas, but also when deciding where to place area midpoints, from which distance, time or cost is measured.

In this chapter we will estimate conventional and new measures of accessibility looking specifically at their ability to handle MAUP-related issues. This is done in two ways. First, in addition to the conventional accessibility models embedding decay functions using flow-data, we introduce a new half-life model, originating from the natural sciences, that uses the median commuting distance (of the entire population). This means that the MAUP stemming from usage of flow-data can be reduced. Secondly, we run all models with both the observed median and mean commuting distance between origin and destination. Differences in accessibility attributable to the use of either median or mean distance indicate that MAUP-related issues can play an unexpected role in the estimation of accessibility.

By introducing new measures and comparing them to the conventional ones, and by contrasting results calculated from median and mean commuting distances, we hope to contribute new knowledge regarding how the spatial organization of jobs and workers in municipalities affect accessibility, and to what extent different methods are more or less sensitive to the MAUP.

The remaining sections are arranged as follows. Section 2.2 describes the adopted methods, while section 2.3 presents the data used in the analysis. Section 2.4 illustrates the results from the accessibility analysis, and finally, section 2.5 provides concluding remarks and discusses the relationship between accessibility, the spatial organization of Swedish municipalities, and the MAUP.

2.2 THE MODELS

2.2.1 The Potential Accessibility

The concept of accessibility came to the fore in the 1950s, thanks essentially to Hansen (1959) who defined accessibility as the potential of opportunities for interactions (Reggiani et al., 2011a). Accessibility can be interpreted as the ease with which activities can be reached from a certain location (Morris et al., 1978) by identifying the interrelationship between the performance of the transportation network and the spatial structure of land use. If we interpret accessibility as the spatial arrangement of employment opportunities (see also Östh, 2007), it is important to investigate whether observed accessibility patterns, which occur as a consequence of the spatial organization of aggregate homes and jobs, exhibit high or low levels of homogeneity.

It is well known from the literature that accessibility as the potential of opportunities can be formulated as follows:

$$Acc_i = \sum_j D_j f(\beta, d_{ij}) \tag{2.1}$$

where Acc_i defines the accessibility of zone i; D_j is a measure (or weight) of opportunities/activities in j; and $f(\beta, d_{ij})$ is the impedance function measuring separation effects between origin i and destination j, where d_{ij} is the (generalized) cost/distance between i and j, and β its cost–distance sensitivity parameter. Interestingly, equation (2.1) also emerges as the inverse of the calibration factor in an origin-constrained spatial interaction model (SIM).

Equation (2.1) clearly shows how important a very good estimate of the β parameter is when constructing suitable accessibility indicators of the spatial landscape under analysis, given D_j and d_{ij}.

This leads us to the following research question: which model provides the best fit for the β parameter? Given the fact that we are in the absence of real data[1] concerning Acc_i, our first step is to explore the goodness of the β parameters emerging from the spatial flows that can be modelled using the rich family of spatial interaction models (SIMs).

The first empirical step of our accessibility analysis is therefore the modelling of spatial flows using different SIMs to extract the cost–distance sensitivity parameters β. These β parameters will be compared in their fitness, and then introduced in equation (2.1) to construct the accessibility measures for the Swedish municipalities.

2.2.2 The Spatial Interaction Model

Preface

Spatial interaction models (SIMs) are static models aiming to describe and predict the analysis of spatial movements, specifically, the processes or spatial flows resulting from given spatial configurations. SIMs are thus models of spatial flows, that is flows of people, commodities, capital, information, and so on, from some origin i to some destination j. SIMs gained a lot of popularity in the past for their usefulness in studying the geography of movement. Nowadays SIMs are still considered a relevant technique for exploring the cohesion and dispersion of activities in spatial systems (Reggiani, 2012, 2014).

SIMs have a long history, starting from analogies with Newton's law of gravity (Wilson, 1981). Subsequently SIMs emerged from two major theories: (1) the entropy theory; and (2) the utility maximizing approach. Consequently, SIMs can be interpreted in a behavioural context with an economic meaning as aggregate models of human behaviour. Interestingly, SIMs are also the basis for modelling accessibility, as it will be subsequently highlighted.

The unconstrained spatial interaction model

A SIM – in its unconstrained form – reads as follows:

$$T_{ij} = K \cdot O_i \cdot D_j \cdot f(\beta, d_{ij}), \tag{2.2}$$

where T_{ij} represent the number of flows which move, communicate or interact (physically or virtually) from the origin i to the destination j. They are a function of the outflows O_i and of the inflows D_j, as well as of the impedance function $f(\beta, d_{ij})$; d_{ij} is the (generalized) cost or the distance between i and j, and the parameter K is a scaling factor, which has to be calibrated. It should be noted that (aggregate) behaviour is embedded in $f(\beta, d_{ij})$ and identified by means of its β-value. Thus the β-value emerging from SIM (2.2) will be the focus of our analysis and will also be used for constructing accessibility indicators according to equation (2.1).

Interestingly, it is worth pointing out the emerging theoretical strength of accessibility when derived from SIMs. In particular (Reggiani, 2012):

- Accessibility (emerging from SIMs) can be linked to statistical information principles and entropy maximization.
- Accessibility (emerging from SIMs) can be linked to logit models and thus to microeconomic theory (stochastic utility maximization).

In this framework it then appears relevant to explore a more powerful SIM, the doubly constrained SIM.

The doubly constrained spatial interaction model

The general form of a (doubly constrained) SIM reads as follows:

$$T_{ij} = A_i B_j O_i D_j f(\beta, d_{ij}) \; i = 1, \ldots, I; j = 1, \ldots, J \tag{2.3}$$

where A_i and B_j are known as balancing factors, equal to:

$$A_i = 1/\sum_j B_j D_j f(\beta, d_{ij}); \; B_j = 1/\sum_i A_i O_i f(\beta, d_{ij}), \tag{2.4}$$

derived from the respective additivity conditions:

$$\sum_j T_{ij} = O_i; \sum_i T_{ij} = D_j \tag{2.5}$$

The variables in equation (2.3) are the same as in equation (2.2). Also equation (2.3) is compatible with Newton's law of gravity. However, it can also emerge from a probabilistic approach based on statistical equilibrium concepts, as proposed by Wilson (1970, 1981). Wilson in particular demonstrated that the type (2.3) SIM can be derived by maximizing an entropy function and can thus be seen as an optimum system solution. In particular, the entropy maximizing approach, by means of the constraints (2.5), provides the factors A_i and B_j in (2.3) (instead of simply K); in addition, it provides the form of the impedance function $f(\beta, d_{ij})$, by means of the following constraint on the total distance d^*:

$$\sum_{ij} d_{ij} T_{ij} = d^* \tag{2.6}$$

The entropy approach offered a macro-behavioural context to the type (2.3) SIM, given the fact that entropy can be interpreted in terms of a generalized cost function for spatial interaction behaviour (Nijkamp, 1975). In addition, microeconomic (deterministic or stochastic) choice theory can offer a more interesting behavioural interpretation to the type (2.3) SIM (Reggiani, 2012). This behavioural framework undoubtedly outlines the relevance and strength of the type (2.3) SIM – and thus of the associated Acc_i (2.1) – in spatial economic science.

Modelling job accessibility

Besides choosing geographically and statistically sound data for the analysis, designing a study of potential accessibility also means deciding which SIM model is the most suitable according to different spatial economic settings and constraints. In most cases, the choices are intuitive: an unconstrained type (2.2) SIM is preferable if there are no constraints in either supply or demand. If, on the other hand, both the supply and demand sides are constrained, a doubly constrained SIM, as in equation (2.3), is preferable.[2] The labour market is the conventional example of the doubly constrained SIM: an employee cannot be working at several locations and with different tasks at the same time, and he/she cannot occupy already occupied jobs (that is both the supply side and the demand side are fixed). Since we are modelling potential accessibility to jobs in the labour market, a doubly constrained SIM is being applied in our experiments to explore accessibility in Sweden. In addition, we also apply an unconstrained type (2.2) SIM (see section 2.4). The latter is used to test to what extent a choice of a less appropriate model affects the outcome, mostly in reference to accessibility ranking. The results are described in section 2.4.

2.2.3 Modelling Distance Decay

Exponential and power decay in spatial interaction models

In addition to the question of the 'best' β-value to be introduced in the formulation of accessibility (2.1), a second issue concerns which type of impedance function better fits the connectivity/economic structure associated with the system under analysis.

In this context, recent methodological experiments on SIMs focused on two different functional specifications of $f(\beta, d_{ij})$, the exponential decay function:

$$f(d_{ij}) = e^{-\beta d_{ij}}, \tag{2.7}$$

and the power decay function:[3]

$$f(d_{ij}) = d_{ij}^{-\gamma}, \tag{2.8}$$

where the coefficients β and γ represent the distance-sensitivity parameters.

These two functions are known for their capability to embed homogeneous versus heterogeneous patterns, respectively, according to De Vries et al. (2009), Richardson (1969) and Willigers et al. (2007). In particular, the exponential decay (2.7) is consistent with the assumption of a constant distance decay parameter for all trip makers, who are then homogeneous with respect to this dimension. On the other hand, the power decay function (2.8) is consistent with a gamma distribution for the distance decay parameter, that is, the population of trip makers is heterogeneous with respect to this parameter (Fotheringham and O'Kelly, 1989; Reggiani et al., 2011b).

Consequently, each unconstrained type (2.2) SIM, or doubly constrained type (2.3), can embed in principle either the exponential decay function (2.7) or the power decay function (2.8). Obviously other different functions can be adopted as well, such as the exponential normal decay function, the exponential square root decay function, the log-normal decay function, and so on (see, e.g., Olsson, 1980; Taylor, 1971). However, forms (2.7) and (2.8) appear to be the most popular, given also their theoretical roots connected to the constraint (2.6) in the entropy maximizing approach (see, e.g., Reggiani et al., 2011b). In particular, the power decay function is found to be suitable for long-distance interactions, since power decay shows a higher tail than the exponential one.

In our empirical application, devoted to analysing the commuting patterns in Swedish municipalities, the unconstrained SIM (2.2) and the doubly constrained SIM (2.3) will be embedded in both forms (2.7) and (2.8) to extrapolate the different β- and γ-values that will subsequently be used in the accessibility function (2.1). The ranking of these four different accessibility indicators will allow us to determine the most suitable SIM model, as well as the associated connectivity patterns (section 2.4).

Applying half-life calculations to distance decay functions

In physics, biology, chemistry and many other scientific fields, half-life models have long been used to estimate the decay of various substances over time. The 'conventional' use of this model calculates the decay of the radioactive isotope Carbon-14, commonly used to date organic material. Radioactive decay in Carbon-14, as for most other isotopes, happens exponentially, which is why the commonly used decay model is identical to the exponential decay function formulated in equation (2.7). Half-life models differ from conventional SIMs in how the β-value is derived and in what fields of research it is being used. In equation (2.9), the formulation for retrieving the β-value is expressed as follows:

$$-\beta = \frac{\ln(0.5)}{m}, \tag{2.9}$$

where m equals the time to half-life of the measured isotope. Translating this model from the rates of decay of radioactive isotopes to an accessibility model describing access to jobs means that variables representing years and radioactivity need to be replaced with more appropriate factors. First, the time variable (in years) can be substituted with a variable describing commuting distance. Since the available Swedish data allows for computation of Cartesian distances between home and job, but is lacking corresponding variables for commuting times or modes of transportation, the variable chosen to represent commuting distance is the Cartesian distance in metres. As a replacement for decay of radioactivity, accessibility to jobs is chosen. In addition, we also need to specify what commuted distance might function as a candidate for replacing the value of the time span between no decay and half-decay in Carbon-14.

The observed median commuting distance is a viable candidate for replacing time to half-life of radioactive decay. The reason being that the median commuting distance partitions the studied population into two equally large groups (just as the cut-off point of 5740 years partitions decay into two equally large groups). If we think of the process of decay as an integral – the AUC (area under the curve) should equal ½ at the median commuting distance. Since the median distance can be used to construct a decay parameter (β; see equation 2.9) that forces half of the AUC to take place exactly at the median distance, it is possible to estimate accessibility to jobs using a model in which half of all available accessibility to jobs happens on each side of the half-life divide of median commuting distance.

In Figure 2A.1 (see Appendix) a hypothetical model illustrates the relationship between decay in accessibility to any one job at various distances. In the example a median commuting value of 5000 metres has been selected for the calculation of the decay parameter (β). A decay parameter value of 0.000138629/m has been calculated.[4] The relationship between access and AUC, as expressed in Figure 2A.1, reveals that the half-life specified β-value 'forces' access (bars) and cumulative AUC (line) to intersect at value 0.5. In reality, the spatial organization of jobs and homes, and the heterogeneity – both in terms of the kinds of skills that are needed at different locations and what kind of individual time–geographical restrictions are at play for different groups of commuters – will make it hard or even impossible to illustrate job accessibility decay as in Figure 2A.1. The underlying idea is nevertheless the same: observed median distance represents the distance within which 50 per cent of the commuters are occupied at the same time as 50 per cent of the commuters need to travel further for employment. And if we perceive the AUC as the average sum of job access and use a parameter that intersects 50 per cent of the job accessibility with the median commuted distance, then we apply a model where, theoretically speaking, equal parts of the opportunities in the labour market are found on either side of the median commuting distance threshold.

Since half-life models are not commonplace in spatial interaction literature, a mathematical basis for the construction of the β-value is necessary.[5] The first step is to perceive of decay as an integral. The total AUC (AUC in an integral function) can be understood as the sum of access to a single job over distance. This total area can be formulated mathematically as an integral (equation 2.10):

$$\int_0^\infty e^{-\beta x} dx = 1/\beta, \tag{2.10}$$

where ∫ represents the integral between time zero (0) and eternity ∞, $e^{\beta x}$ represents the exponential function, and finally $dx = 1/\beta$ represents the integrated area. Since the distance to the 'half-life' of accessibility coincides with half of the AUC, the integral for half-life and half-AUC can be formulated as in equation (2.11) or (2.12):

$$\int_0^m e^{-\beta x} dx = 0.5/\beta \tag{2.11}$$

$$0.5 = \beta \int_0^m e^{\beta x} dx = 1 - e^{-\beta m} \tag{2.12}$$

The differences between equation (2.10) and equations (2.11) and (2.12) consist of changes in the span of the integral from time zero to time m, as well as a reduction of the integrated area from 1 to 0.5. In this example the distance m represents the 5000 metres representing median commuting distance. The remaining unknown value is the parameter (β) which can be determined by rewriting equation (2.12) as in equation (2.13) below:

$$0.5 = e^{-\beta m} \tag{2.13}$$

Taking natural logs (ln):

$$\ln(0.5) = -\beta m \tag{2.14}$$

And finally solving for β, as in equation (2.15):

$$-\beta = \frac{\ln(0.5)}{m} \tag{2.15}$$

How then does the half-life model compare to unconstrained and doubly constrained SIMs in terms of producing parameters for decay? An important difference between the two groups is the fact that half-life decay parameters are derived mathematically, while the conventional parameters, or those emerging from SIMs, are produced statistically. This means that conventional parameters can be evaluated according to how well they fit statistically, with reference to estimated flows versus observed flows. The benefit of using half-life models is that decay parameters can be produced also when the MAUP is considered to be substantial, or when flow data is missing or incomplete. This is because the parameter is constructed using data on a national (or any other) level.

The role of the mean versus the median value of the distance

The flow data used in this chapter is constructed from data describing residential and work coordinates for each working individual in Sweden. This means that the Cartesian distance between home and job for each individual can be expressed using Pythagoras' theorem. The distance commuted between nodes is therefore not limited to a generalized measure between municipality midpoints, but can be expanded to encompass median and mean distances.

Choosing between midpoint distances and/or mean and median distances may seem trivial, but it is a choice that may affect accessibility estimates to a great extent. That cross-border commutes happen more often when municipalities are adjacent or closely located makes intuitive sense. Consequently, using the observed median commuting distance may be a good way of capturing local commuting behaviour. Following the same argumentation, mean commuting distances would be exaggerating the lengths of the typical cross-border commuting distance. All else being equal, models that are less sensitive to the types of distance used should be favoured if knowledge about the spatial organization of jobs and homes within the studied areas is scarce.

2.3 DATA AND METHODS

2.3.1 Source and Use of Data

All data used in this study comes from the Statistics Sweden (SCB) PLACE database located at the Department of Social and Economic Geography at Uppsala University. PLACE currently contains statistics on all Sweden resident individuals between 1990 and 2008. The database contains longitudinal statistics on individuals' education, employment, demography and geography. Since the geographic resolution of both residential and workplace locations is high, it is possible not only to calculate the flow of individuals between municipalities, but also to accurately describe mean and median distances commuted both within and between municipalities.

2.3.2 Choice of Years

In order to test our hypotheses we designed a dataset with commuting patterns from 1993 and compared it to the commuting patterns in 2008. The year 1993 was chosen as the first reference point for two reasons. Firstly, the quality of the data is considerably better in 1993, especially when compared to 1990, making 1993 a more suitable reference year. Secondly, the Swedish economy changed considerably during the early 1990s, which makes the period less suitable as a starting point for the analysis (Östh and Lindgren, 2012).

From the late 1980s to 1992 the Swedish economy experienced an economic boom followed by a deep recession. In 1993 the economy started to recover, but more importantly, the economy began to move from an industry-based economy to today's service-oriented and globalized economy. We chose 2008 as the last year of study since 2008 currently represents the most recent year available in the database. Due to the above-listed restrictions, changes in job accessibility can be studied for a 15-year interval.

2.3.3 Choice of Geographical Units

Sweden can be described as a sparsely populated country with small polycentric structures. Most Swedish municipalities consist of small to mid-sized towns with surrounding rural areas, with the exception of municipalities in the greater Stockholm region where some lack rural surroundings (Johansson, 2002). Compared to other geographical units (counties/NUTS 3 regions are too large, and parishes and small area market statistics or SAMS are too small) the administrative size of municipalities is ideal for describing labour market accessibility.

In order to construct a dataset that can be used to compare accessibility, the geographic parameters of the dataset need to be tweaked to make sense. These adjustments are necessary since in some cases administrative borders have been changed or municipalities have merged or been separated. Unless these changes are taken into account the comparison can be meaningless. A study of changing accessibility patterns in Sweden over time is no exception. The official count of municipalities increased from 286 in 1993 to 290 in 2008, making direct comparisons between municipalities inappropriate. To solve this problem and make sure that the accessibility estimates were constructed using the same set of spatial units, we used a Geographic Information System (GIS) tool where the observed coordinates of work and residence for each employed individual were used to place the 1993 data within the administrative borders of 2008.

2.3.4 The Commuting Population

The total count of jobs, workers and flows between and within any set of municipalities is based on the annual registers available in PLACE. The aggregate municipality-level data is composed from a dataset containing all individuals registered as working during a measurement week in November in either 1993 or 2008. The coordinates are used to calculate Cartesian distances between home and work for each worker, which in turn are being used to denote commuting distances. Cartesian distances rather than travel time or travel costs are used for commuting because the database does not contain information on either modes or routes of transportation.

2.3.5 The Flow Matrixes

The flow of commuters between and within municipalities is aggregated into two datasets, one for 1993 and one for 2008. The two matrixes containing the flow of people between any municipality i and j was created through aggregation of micro-level data, as specified above. Since all 290 municipalities can potentially interact with each other the aggregated datasets contain 290 * 290 possible combinations. However, not even the most populous of all Swedish municipalities (Stockholm) is recorded to have commuters to every other Swedish municipality. In 1993 the average municipality had commuter flows to 170 others; in 2008 that number had increased to 189. Variables in the matrixes include municipality of origin i and municipality of destination j, the flow of commuters between i and j and the observed median and mean commuting distances between i and j.

2.4 THE EMPIRICAL ANALYSIS: ACCESSIBILITY IN SWEDEN

2.4.1 The Estimation Results from SIMs

As anticipated in section 2.2, we will analyse the following accessibility formulations on the basis of the potential accessibility (2.1):

$$Acc_i exp = \sum_j D_j (e^{-\beta d_{ij}}), \qquad (2.16)$$

$$Acc_i pow = \sum_j D_j (d_{ij}^{-\gamma}), \qquad (2.17)$$

where the coefficients β and γ represent the distance-sensitivity parameters. In particular, the coefficients β and γ have been estimated from the calibration procedure[6] concerning the unconstrained SIM (2.2), as well as the doubly constrained SIM (2.3), by taking into account: (1) the two aforementioned impedance functions (exponential and power form); and (2) the mean and median distance. In addition, the β-value of the half-life model has also been introduced in equation (2.16). It should also be noted that the logarithmic expression of equation (2.16), the so-called logsum measure, which is in agreement with consumer theory (Ben-Akiva and Lerman, 1985), has been often used in accessibility studies (see, e.g., Geurs et al., 2012; Nicolai and Nagel, 2012).

For either mean or median distances, and for both 1993 and 2008, we have computed four accessibility indicators based on the parameters of two SIMs, plus the accessibility indicator (2.16) calculated from the half-life parameter. All these accessibility models have then been compared in order to highlight the most accessible municipalities in Sweden, and thus to extrapolate some considerations on the underlying spatial economic network structure (see next section 2.4.2). The calibration procedure concerning SIM (2.2) and (2.3) has led to the results illustrated in Table 2.1.

From Table 2.1, we can outline the following findings:

Table 2.1 Calibration results concerning unconstrained SIM and doubly constrained SIM; commuting flows between 290 Swedish municipalities, 1993 and 2008

		Unconstrained				Doubly constrained			
		Exponential decay mean	Exponential decay median	Power decay mean	Power decay median	Exponential decay mean	Exponential decay median	Power decay mean	Power decay median
1993	Parameters	−0.003642009	−0.00365	−1.32739	−1.30561	0.058237	0.057438	1.896429	1.643166
	R^2	0.215	0.215	0.558	0.577	0.950	0.955	0.993	0.993
	t-value	−88.661	−88.807	−190.384	−198.013	94.653	94.128	96.898	114.234
2008	Parameters	−0.003572233	−0.00358	−1.3402	−1.31782	0.047407	0.047005	1.830314	1.587367
	R^2	0.221	0.222	0.558	0.574	0.956	0.96	0.991	0.99
	t-value	−100.591	−100.758	−212.009	−219.04	107.464	107.302	139.587	147.885

- The doubly constrained SIM always fits better than the unconstrained SIM, as expected.
- Both SIMs embedding power decay functions always fit better than those embedding exponential decay functions.
- In the doubly constrained SIM, the power decay function fits better than the exponential decay function, although the difference is only around 4 per cent.
- The inclusion of either the mean or the median in both SIMs does not change the R^2 values.

In summary, on the basis of the better R^2 values concerning the fitting of the power impedance function, we can conjecture an associated commuting network with a tendency towards polycentrism (Reggiani et al., 2011a, 20011b). This hypothesis will be tested by visualizing all the accessibility indicators computed from all of the models used in the study.

2.4.2 The Most Accessible Municipalities in Sweden

In this section we show the results of the calculation of the two main accessibility indicators (2.16) and (2.17), where the parameters β and γ have been extracted from the calibration of SIMs (2.2) and (2.3). These two accessibilities embed the decay functions (2.7) and (2.8), in which either the mean value or the median value of the distance is considered. To these accessibility models we add the accessibility model (2.16) computed from the half-life decay function (2.14) and estimate accessibility for 1993 and 2008. The results are presented as maps, as well as tables, illustrating accessibility rankings.

In order to point out the between-model deviations, the accessibility ranking for each of the models is presented in Figure 2.1, which contains graphics illustrating accessibility in 2008[7] and is arranged as follows: the unconstrained models are presented in the top row, and the doubly constrained models in the bottom row. The left-most section displays power models, the middle section exponential models, and finally the right-most section displays the half-life models. Mean and median distance accessibility ranks are presented for each of the models. The general pattern recognizable in all of the models is that accessibility is greater in the most populous Swedish areas. The power models are more restrained in terms of the spread of higher accessibility outside the populous core areas in both the doubly and unconstrained models. This difference can be seen when compared to the exponential model, but is even more evident for the half-life model. Similarly, higher accessibility is concentrated in more populous regions using doubly constrained models compared to unconstrained models.

Although the maps cannot statistically prove the difference between

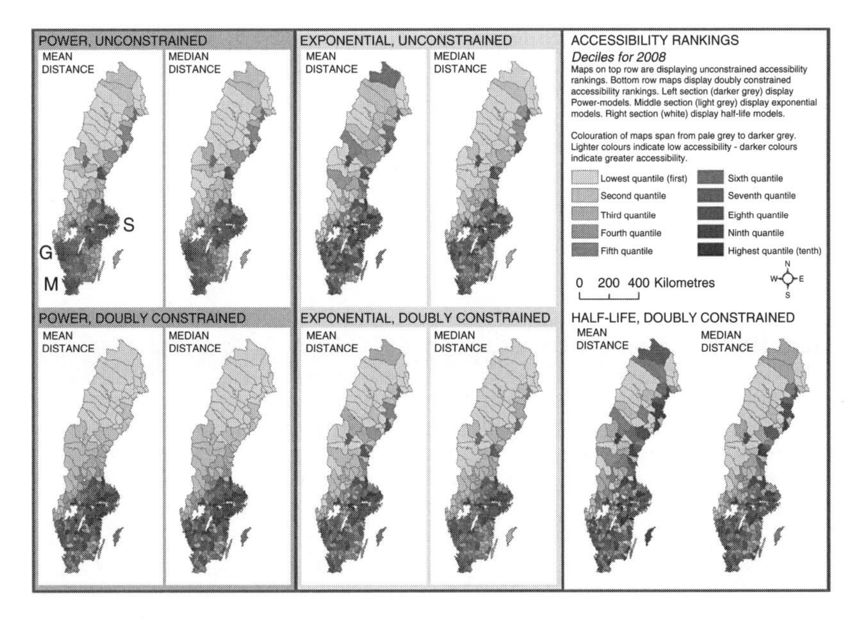

Note: Letters S, G and M, in the upper left map point out the location of Stockholm (S), Göteborg (G) and Malmö (M).

Figure 2.1 Ten different accessibility rankings for the 290 municipalities in Sweden, 2008

models, how the distances are defined (mean or median) seems to be important. Exponential accessibility models (associated to both the SIM and half-life approach) seem to be especially sensitive to this, while power accessibility models are affected to a much lesser extent.

Table 2.2 and Table 2A.1 (in the Appendix) represent the ranking of accessibility from the highest accessibility on the top to lowest accessibility at the bottom (note that only the top 15 and bottom 10 are included) for 2008 (Table 2.2) and 1993 (Table 2A.1).

The comparison, both between years and between models, reveals the following: (1) the output is similar for all types of models, including the unconstrained SIMs, at least for the top-ranked cities: in particular, the results for the power-unconstrained SIM perfectly echo the top-five ranking of the doubly constrained SIM; and (2) the results closely match the ranked distribution of jobs on the Swedish labour market. Municipalities in the Stockholm region (dark grey) are ranked as having the highest accessibility; some municipalities in the Göteborg area (middle grey) and Malmö area (this includes the municipalities of Malmö and

Table 2.2 Accessibility of the first 15 and last 10 Swedish municipalities for 2008; commuting network between the 290 Swedish municipalities

Mean distance				
Exponential decay			Power decay	
Doubly constrained SIM	Unconstrained SIM	Half-life*	Doubly constrained SIM	Unconstrained SIM
Stockholm	Stockholm	Stockholm	Solna	Solna
Solna	Huddinge	Solna	Sundbyberg	Sundbyberg
Sundbyberg	Solna	Sundbyberg	Stockholm	Stockholm
Danderyd	*Linköping*	Danderyd	Danderyd	Danderyd
Lidingö	*Eskilstuna*	Lidingö	Lidingö	Lidingö
Huddinge	*Norrköping*	Huddinge	Malmö	Huddinge
Nacka	Nacka	Sollentuna	Huddinge	Nacka
Sollentuna	Södertälje	Nacka	Sollentuna	Sollentuna
Täby	*Örebro*	Järfälla	Nacka	Göteborg
Järfälla	*Västerås*	Täby	Göteborg	Täby
Ekerö	Botkyrka	Ekerö	Täby	Järfälla
Botkyrka	Sollentuna	Botkyrka	Järfälla	Malmö
Tyresö	Järfälla	Tyresö	Burlöv	Botkyrka
Upplands Väsby	*Uppsala*	Upplands Väsby	Upplands Väsby	Tyresö
Vaxholm	Haninge	Vaxholm	Botkyrka	Upplands Väsby
. . .	. . .	. . .	. . .	. . .
Överkalix	Malå	Överkalix	Överkalix	Malå
Övertorneå	Haparanda	Åsele	Åsele	Dorotea
Dorotea	Sorsele	Övertorneå	Vilhelmina	Åsele
Åsele	Jokkmokk	Dorotea	Dorotea	Storuman
Vilhelmina	Gällivare	Jokkmokk	Övertorneå	Överkalix
Jokkmokk	Arjeplog	Vilhelmina	Jokkmokk	Jokkmokk
Storuman	Övertorneå	Sorsele	Arjeplog	Övertorneå
Pajala	Överkalix	Storuman	Storuman	Arjeplog
Arjeplog	Kiruna	Pajala	Pajala	Sorsele
Sorsele	Pajala	Arjeplog	Sorsele	Pajala

Note: * beta value = 0.000029243/m; ** beta value = 0.00011613911264/m.

Burlöv) are often also listed as high accessibility municipalities. Outside of the Stockholm, Göteborg and Malmö regions (white with text in italics), and with one exception, the rest of the municipalities are ranked as highly accessible only by unconstrained exponential models.

It should be noted that the unconstrained exponential models pick up larger municipalities (in population size ranked directly below Stockholm, Göteborg and Malmö) as having high ranks of accessibility but apart from

Median distance				
Exponential decay			Power decay	
Doubly constrained SIM	Unconstrained SIM	Half-life**	Doubly constrained SIM	Unconstrained SIM
Stockholm	Stockholm	Stockholm	Solna	Solna
Solna	Huddinge	Solna	Sundbyberg	Stockholm
Sundbyberg	Solna	Sundbyberg	Stockholm	Sundbyberg
Lidingö	*Linköping*	Lidingö	Danderyd	Danderyd
Danderyd	*Eskilstuna*	Danderyd	Lidingö	Lidingö
Huddinge	*Norrköping*	Huddinge	Huddinge	Huddinge
Nacka	Nacka	Nacka	Nacka	Nacka
Sollentuna	Södertälje	Sollentuna	*Gnesta*	Sollentuna
Täby	*Örebro*	Göteborg	Göteborg	Göteborg
Järfälla	Botkyrka	Täby	Sollentuna	Malmö
Ekerö	Sollentuna	Järfälla	Malmö	Täby
Botkyrka	*Västerås*	Tyresö	Täby	Järfälla
Tyresö	Järfälla	Botkyrka	Järfälla	Botkyrka
Upplands Väsby	*Uppsala*	Ekerö	Botkyrka	Tyresö
Haninge	Haninge	Mölndal	*Lund*	Mölndal
...	...	...	...	...
Överkalix	Malå	Jokkmokk	Älvdalen	Ragunda
Vilhelmina	Haparanda	Storuman	Pajala	Malå
Dorotea	Sorsele	Pajala	Arjeplog	Åsele
Övertorneå	Jokkmokk	Övertorneå	Dorotea	Dorotea
Jokkmokk	Gällivare	Bjurholm	Övertorneå	Pajala
Storuman	Arjeplog	Arjeplog	Ragunda	Arjeplog
Åsele	Övertorneå	Överkalix	Berg	Övertorneå
Pajala	Överkalix	Dorotea	Storuman	Storuman
Sorsele	Kiruna	Åsele	Sorsele	Sorsele
Arjeplog	Pajala	Sorsele	Överkalix	Överkalix

this apparent difference it is interesting to note how similar the results are for unconstrained and doubly constrained ranking. Both unconstrained and doubly constrained power models rank municipalities in the three metropolitan areas as most accessible. Regardless of year and model the municipalities ranked as having the poorest accessibility are all located in rural areas of Sweden, far from any of the metropolitan areas. It should be noted that commuting flows and the number of jobs in adjacent municipalities in neighbouring countries are not included in the models due to data restrictions. This will likely affect estimates from unconstrained models more than constrained models. In addition, municipalities like Malmö being located close to Copenhagen will be more affected than other more distant municipalities.

If we compare the ranking tables for differences when using either the mean or median commuting distance, changes in accessibility rank can be found. The changes are small (gaining or losing a few positions at most when comparing corresponding SIMs), but indicate that the method used for measuring commuting within and between municipalities has an impact on the overall results.

2.5 CONCLUSIONS

In this chapter we have analysed conventional and new approaches to the estimation of accessibility using Swedish data from 1993 and 2008. The purpose of our study has been to analyse the accessibility patterns in Sweden over a 15-year period. A second underlying goal of our accessibility analysis was to test whether the results are affected by the spatial organization of the units of analysis or the way distance is measured.

The comparison of accessibility between 1993 and 2008 indicates that the pattern of accessibility is stable over time. The accessibility ranking derived from different spatial interaction models reveals that most municipalities will gain or lose only a few positions in the ranking regardless of what model, year or representation of distance we use. The stability of accessibility ranking can partly be accredited to the equal distribution of infrastructure improvements (commuting flows have increased substantially in all of Sweden over time), but also to the spatial organization of Swedish municipalities, which can be understood as spatial units where most people not only work but also reside. Our analysis of the effects that the MAUP has on accessibility at the municipality level in Sweden revealed that how commuting distances are measured within and between municipalities has an effect on accessibility. The result is not surprising, but it is a relevant reminder that the way spatial data is gathered is of importance for analysis. The least sensitive model to the MAUP seems to be the power accessibility model, emerging from both the unconstrained and doubly constrained SIMs, since the results change only slightly when comparing mean and median commuting distance. Consequently, the power model is to be preferred if the quality of the flow data is uncertain. The half-life model showed findings that were similar to our exponential accessibility results emerging from the SIM, indicating that half-life models can be used as a second-best alternative. This means that when knowledge of commuting flows and distances is confined to national or large-scale regional levels, or when spatial interaction analysis is conducted on a very disaggregated or even individual level, half-life models can be viable candidates for the production of decay parameters.

NOTES

1. Accessibility is not measurable, thus calibration methods can be carried out only with reference to the transport flows, which are instead measurable.
2. Singly constrained SIMs should be applied if either the demand or supply side is fixed.
3. The power decay function can also be formulated as: $f(d_{ij}) = e^{-\gamma \ln(d_{ij})}$. Using a logarithmic formulation more clearly shows that the disutility of travel decays more quickly with short distances compared to longer distances, meaning that an extra kilometre added to a short trip is perceived as greater than an extra kilometre added to a long trip (Wilson, 1981; Illenberger et al., 2012).
4. The value of the parameter may vary depending on the metric system used for measuring distance. Using kilometres renders 0.138629436/km (median commuting distance = 5 km), miles renders 0.223102451/mi (median commuting distance = 3.10685596 miles); both of these alternative constant values will produce identical outcomes as long as appropriate metric systems are used.
5. We have not found any references suggesting that half-life models are being or have been used in spatial interaction modelling.
6. The calibration procedure concerning the unconstrained SIM is based on a simple regression analysis; see, for details, Reggiani et al. (2011a). The calibration procedure concerning the doubly constrained SIM is based on an iterative procedure, as indicated by Wilson (1981).
7. Maps illustrating accessibility for 1993 are similar and have been left out due to space restrictions.

REFERENCES

Ben-Akiva, M. and S.R. Lerman (1985), *Discrete Choice Analysis*, Cambridge, MA: MIT Press.

De Vries, J., P. Nijkamp and P. Rietveld (2009), ‘Exponential or power distance-decay for commuting? An alternative specification’, *Environment and Planning A*, **41** (2), 461–480.

Fotheringham, A.S. and M.E. O’Kelly (1989), *Spatial Interaction Models: Formulations and Applications*, Dordrecht: Kluwer Academic.

Fotheringham, A.S. and D.W.S. Wong (1991), ‘The modifiable areal unit problem in multivariate statistical analysis’, *Environment and Planning A*, **23** (7), 1025–1044.

Geurs, K.T., M. de Bok and B. Zondag (2012), ‘Accessibility benefits of integrated land use and public transport policy plans in the Netherlands’, in K.T. Geurs, K.J. Krizek and A. Reggiani (eds), *Accessibility and Transport Planning: Challenges for Europe and North America*, Cheltenham, UK and Northampton, MA, USA: Edward Elgar, pp. 135–153.

Handy, S.L. and D.A. Niemeier (1997), ‘Measuring accessibility: an exploration of issues and alternatives’, *Environment and Planning A*, **29**, 1175–1194.

Hansen, W. (1959), 'How accessibility shapes land use', *Journal of the American Institute of Planners*, **25**, 73–76.
Illenberger, A.J., K. Nagel and G. Flötteröd (2012), 'The role of spatial interaction in social networks', *Networks and Spatial Economics*, **13** (3), 255–282.
Johansson, M. (2002), 'Polycentric urban structures in Sweden – conditions and prospects', in C. Bengs (ed.), *Facing ESPON*, Nordregio Report 2002:1, Stockholm, http://www.nordregio.se/Global/Publications/Publications%20 2002/R2002_1/R0201_p99.pdf.
Kwan, M-P. (1998), 'Space–time and integral measures of individual accessibility: a comparative analysis using a point-based framework', *Geographical Analysis*, **30** (3), 191–216.
Morris, J.M., P.L. Dumble and M.R. Wigan (1978), 'Accessibility indicators for transport planning', *Transportation Research A*, **13**, 91–109.
Nicolai, T.W. and K. Nagel (2012), 'Coupling transport and land-use: investigating accessibility indicators for feedback from a travel to a land use model', https://svn.vsp.tu-berlin.de/repos/public-svn/publications/vspwp/2012/12–16/2012–08–03_accessibility_accepted_latsis.pdf.
Nijkamp, P. (1975), 'Reflections on gravity and entropy models', *Regional Science and Urban Economics*, **5**, 203–225.
Olsson, G. (1980), *Birds in Egg/Eggs in Bird*, London: Pion.
Openshaw, S. (1984), *The Modifiable Areal Unit Problem*, Norwich: Geo Books.
Östh, J. (2007), *Home, Job and Space: Mapping and Modeling the Labor Market*, Geografiska Regionstudier 72, Uppsala: Uppsala University.
Östh, J. (2011), 'Introducing a method for the computation of doubly constrained accessibility models in larger datasets', *Networks and Spatial Economics*, **11**, 581–620.
Östh, J. and U. Lindgren (2012), 'Do changes in GDP influence commuting distances? A study of Swedish commuting patterns between 1990 and 2006', *Tijdschrift Volume voor Economische en Sociale Geografie*, **103** (4), 443–456.
Reggiani, A. (2012), 'Accessibility, connectivity and resilience in complex networks', in K.T. Geurs, K.L. Krizek and A Reggiani (eds), *Accessibility and Transport Planning: Challenges for Europe and North America*, Cheltenham, UK and Northampton, MA, USA: Edward Elgar, pp. 15–36.
Reggiani, A. (2014), 'Complexity and spatial networks', in M. Fischer and P. Nijkamp (eds), *Handbook of Regional Science*, Berlin and Heidelberg, Germany and New York, USA: Springer, pp. 811–832.
Reggiani, A., P. Bucci, and G. Russo (2011a), 'Accessibility and network structures in the German commuting', *Networks and Spatial Economics*, **11**, 621–641.
Reggiani, A., P. Bucci and G. Russo (2011b), 'Accessibility and impedance forms: empirical applications to the German commuting networks', *International Regional Science Review*, **34** (2), 230–252.
Richardson, H.W. (1969), *Elements of Regional Economics*, Harmondsworth: Penguin Books.
Taylor, P.J. (1971), 'Distance transformation and distance decay functions', *Geographical Analysis*, **3**, 221–238.
Willigers, J., H. Floor and B. Van Wee (2007), 'Accessibility indicators for location choices of offices: an application to the intraregional distributive effects of high-speed rail in the Netherlands', *Environment and Planning A*, **39**, 2086–2098.
Wilson, A.G. (1970), *Entropy in Urban and Regional Modelling*, London: Pion.
Wilson, A.G. (1981), *Geography and the Environment*, Chichester: John Wiley & Sons.

APPENDIX

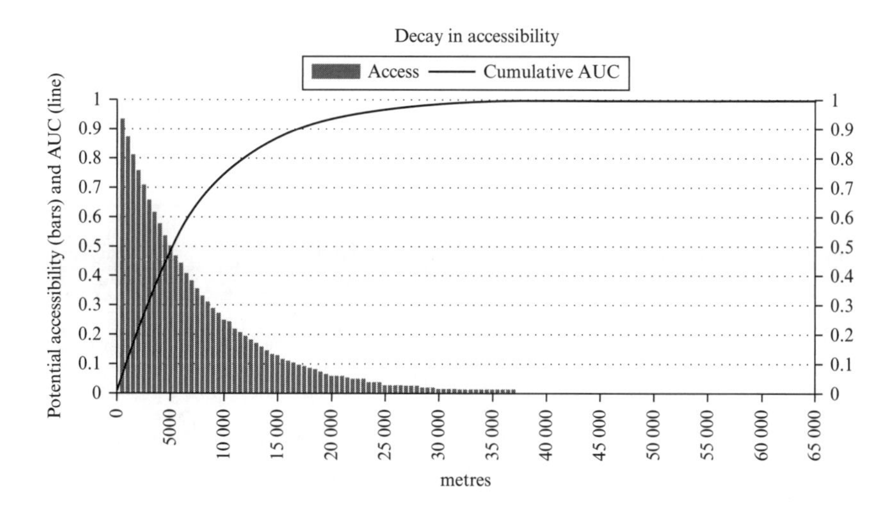

Note: The value 0.5 of cumulative AUC intersects accessibility at value 0.5.

Figure 2A.1 Relationship between access to jobs and cumulative AUC in a hypothetical framework – describing access to any one job at various distances

Table 2A.1 Accessibility of the first 15 and last 10 Swedish municipalities for 1993

Mean distance				
	Exponential decay		Power decay	
Doubly constrained SIM	Unconstrained SIM	Half-life*	Unconstrained SIM	Doubly constrained SIM
Stockholm	Stockholm	Solna	Solna	Solna
Solna	*Linköping*	Stockholm	Sundbyberg	Sundbyberg
Sundbyberg	*Örebro*	Sundbyberg	Stockholm	Stockholm
Danderyd	*Västerås*	Danderyd	Danderyd	Danderyd
Lidingö	Solna	Lidingö	Lidingö	Lidingö
Huddinge	*Uppsala*	Huddinge	Malmö	Huddinge
Sollentuna	Huddinge	Nacka	Huddinge	Sollentuna
Nacka	*Norrköping*	Sollentuna	Göteborg	Nacka
Täby	Södertälje	Täby	Sollentuna	Täby
Järfälla	*Eskilstuna*	Göteborg	Täby	Göteborg
Botkyrka	Nacka	Järfälla	Nacka	Malmö
Tyresö	*Jönköping*	Tyresö	Järfälla	Järfälla
Ekerö	Täby	Botkyrka	Burlöv	Botkyrka
Upplands Väsby	Järfälla	Partille	Upplands Väsby	Tyresö
Vaxholm	Botkyrka	Malmö	Tyresö	Upplands Väsby
...	...	...	...	...
Storuman	Malå	Överkalix	Åsele	Åsele
Malå	Haparanda	Jokkmokk	Vilhelmina	Malå
Jokkmokk	Sorsele	Bjurholm	Överkalix	Dorotea
Övertorneå	Gällivare	Åsele	Jokkmokk	Jokkmokk
Överkalix	Jokkmokk	Övertorneå	Dorotea	Storuman
Pajala	Arjeplog	Dorotea	Övertorneå	Överkalix
Åsele	Kiruna	Pajala	Arjeplog	Övertorneå
Dorotea	Övertorneå	Arjeplog	Storuman	Arjeplog
Sorsele	Överkalix	Storuman	Sorsele	Sorsele
Arjeplog	Pajala	Sorsele	Pajala	Pajala

Note: * beta value = 0.00013186681787/m; ** beta value = 0.00020767435277/m.

Median distance				
Exponential decay			Power decay	
Unconstrained SIM	Doubly constrained SIM	Half-life**	Doubly constrained SIM	Unconstrained SIM
Stockholm	Stockholm	Stockholm	Solna	Solna
Solna	*Linköping*	Solna	Stockholm	Stockholm
Sundbyberg	*Örebro*	Sundbyberg	Sundbyberg	Sundbyberg
Danderyd	*Västerås*	Lidingö	Danderyd	Danderyd
Lidingö	Solna	Danderyd	Lidingö	Lidingö
Huddinge	*Uppsala*	Göteborg	Huddinge	Huddinge
Sollentuna	Huddinge	Huddinge	Malmö	Göteborg
Nacka	Södertälje	Malmö	Göteborg	Sollentuna
Täby	*Norrköping*	Nacka	Sollentuna	Nacka
Järfälla	*Eskilstuna*	Sollentuna	Nacka	Malmö
Botkyrka	Nacka	Täby	Täby	Täby
Tyresö	*Jönköping*	Mölndal	Mölndal	Järfälla
Ekerö	Täby	Partille	*Lund*	Botkyrka
Upplands Väsby	Järfälla	Burlöv	Järfälla	Mölndal
Vaxholm	Botkyrka	Järfälla	Botkyrka	Tyresö
...	...	...	...	...
Storuman	Malå	Ljusnarsberg	Övertorneå	Älvdalen
Malå	Haparanda	Övertorneå	Ragunda	Norsjö
Övertorneå	Sorsele	Ockelbo	Eda	Berg
Jokkmokk	Gällivare	Åsele	Älvdalen	Dorotea
Överkalix	Jokkmokk	Berg	Sorsele	Övertorneå
Pajala	Arjeplog	Överkalix	Arjeplog	Storuman
Åsele	Kiruna	Dorotea	Storuman	Sorsele
Dorotea	Övertorneå	Arjeplog	Berg	Arjeplog
Sorsele	Överkalix	Sorsele	Pajala	Pajala
Arjeplog	Pajala	Bjurholm	Överkalix	Överkalix

3. Transport networks and accessibility: complex spatial interactions

David Philip McArthur, Inge Thorsen and Jan Ubøe

3.1 INTRODUCTION

The concept of accessibility is becoming an increasingly important one when it comes to transportation planning (Straatemeier, 2008). Changes to the transportation network are one of the main causes of changes in accessibility (Mackiewicz and Ratajczak, 1996). Understanding this relationship may help us to plan better transportation networks. Accessibility can be conceptualized at different spatial scales. For instance, we may measure global accessibility by examining airport connections, national accessibility by looking at the motorway network, or urban accessibility by looking at the public transportation network. In this chapter, we concentrate on intra-regional accessibility in a rural to semi-urban context. We are particularly interested in the impact that topographical barriers have on accessibility to the labour market and the effect of removing these barriers.

Accessibility has been used in measuring the accessibility effects of new infrastructure around the world. For example, Holl (2007) examines the accessibility impacts of the Spanish motorway building programme, while Hou and Li (2011) examine the impact of highways and high-speed rail on accessibility in China. Such applications of the accessibility concept raise a number of important questions. One of these questions is exactly how to define and implement accessibility. This is still an area of ongoing research (Reggiani et al., 2011; López et al., 2008; Martín and Reggiani, 2007; Geurs and Van Wee, 2004) and one which is not the central focus of this chapter.

If our aim is to measure accessibility from existing data, then we can choose an accessibility measure which suits our purposes and carry out the necessary calculations. Another potentially useful application of accessibility analysis is in forecasting the impact of an investment in the transportation network. For example, López et al. (2008) use accessibility indicators to predict the impact of infrastructure investment in Spain. This matter can be far more complicated than we might first imagine, since accessibility has been shown to affect location decisions for both firms (de Bok, 2009) and workers (Eliasson et al., 2003). These relocations may reinforce a positive change in accessibility or detract from it. Accounting for the complex interactions between residential and firm location patterns and commuting and migration flows is a key aim of this chapter.

It is our aim in this chapter to use a spatial general equilibrium model to explore changes in accessibility patterns caused by changes in the transportation infrastructure. Of particular interest is the centralization of both the population and economic activity. This is a topic which has long been of interest to researchers. In the new economic geography framework, the degree of centralization in an interregional setting depends on the balance of a number of centrifugal and centripetal forces (Fujita et al., 1999; Krugman, 1991). In an intra-regional setting, Meijers et al. (2012) show that the degree of centralization following changes in the transportation network depends on the spatial structure of the region, the willingness of the population to migrate and commute and the industrial mix.

We use a spatial general equilibrium model to analyse a particular geography at an intra-regional scale. The modelling framework we employ has the advantage of being far simpler than the sort of micro-simulation framework employed by de Bok (2009). Our model captures many of the important features of a more complex and demanding micro-simulation model, in that it allows for spatial interaction in the form of migration, commuting and firm relocation. This is achieved by dispensing with some of the features which are relevant at an urban scale, but which are less important at an intra-regional scale. For instance, route assignment problems are far less relevant in the sort of region we consider, where origins and destinations typically have only one route between them.

The model is used to investigate whether the complex interactions may give rise to unexpected effects when removing a topographical barrier from the transportation network. In Norway, for example, there has been a desire to prevent centralization and promote a geographically dispersed population. One policy tool which is often deployed is improving the connectivity of more peripheral areas. We will explore when such interventions might be effective and when they might achieve the opposite of the intended result.

3.2 A NON-TECHNICAL DESCRIPTION OF THE MODELLING FRAMEWORK

In this section a non-technical description of the model is presented. A technical description can be found in the Appendix. The core of the model centres on the definition of equilibrium, involving intra-regional migration and commuting flows corresponding to a specific spatial distribution of jobs and workers between the zones of a region. To reach an operational model specification we introduce a set of reasonable hypotheses on the spatial behaviour of firms and households. Hence we concentrate on the spatial dimension of the supply and demand for labour.

Consider the demand for labour at specific locations. Like most spatial general equilibrium models, our model incorporates the core elements of economic base modelling (Lowry, 1964; Goldner, 1971). The model therefore distinguishes between two types of firms. The activity of local sector firms is determined by demand arising from within the region, while production in basic sector firms is determined by factors unrelated to intra-regional demand. Total employment is the sum of the employment in both of these sectors (equation 3.1).

Three types of spatial mobility are considered in the model: commuting (equation 3.3), migration (see Appendix sections A2, A3 and equations 3.6–3.10) and shopping behaviour (equations 3.2–3.3).

3.2.1 Basic Sector Firms, Local Innovativity and Competitiveness

The number of basic sector jobs at a specific location depends on local innovativity and competitiveness. This may reflect agglomeration economies, the wage level, entrepreneurial spirit, transport costs, the availability of qualified workers, and so on. There is no attempt to account explicitly for local variations of innovation and competitiveness in the current version of the model. The spatial distribution of basic sector jobs is assumed exogenous in the model.

3.2.2 Local Sector Firms: The Spatial Shopping Pattern

The spatial distribution of local sector jobs reflects the residential location pattern and the spatial shopping behaviour of households (equation 3.3). Retailing is the dominant local sector activity. As in most other forms of spatial interaction there is a distance deterrence effect. This can be used to support an assumption of proportionality between local sector employment and the population of a specific geographic area. It can be argued, however, that such an assumption of proportionality is not very reasonable at the level of spatial aggregation considered in this chapter.

Hence an alternative, more appropriate approach is used. The idea is that consumer shopping behaviour results from a trade-off between price savings and transport costs. Gjestland et al. (2006) derived results for this trade-off. The more realism added to the assumptions about the distribution of price savings, product range, shopping frequency and the valuation of time, the closer Gjestland et al. (2006) come to a smooth function between the frequency of shopping locally and the distance from the shopping centre offering favourable prices.

A next step in deriving a spatial pattern of local sector activities is by recognizing that economies of scale, transportation costs and agglomeration benefits allow firms in a central location to offer goods and services at a lower price than firms located in more peripheral locations. Agglomeration benefits explain why some types of local sector activities will largely be concentrated in a centre. Administrative services often locate in the centre, giving rise to agglomeration benefits which in turn attract more activity. At the same time businesses in many cases choose to locate in the same area because consumers often perceive it to be beneficial if they can satisfy their demand for several goods and services with one shopping trip.

In other words, the potential for price savings pulls shopping towards urban centres, while transport costs contribute to explaining why customers do some of their shopping close to where they live. For an illustration of the outcome of the relevant trade-off, assume that the region has just one centre. Consider next the number of shop employees per resident, measuring the local sector density (LSD). Low prices explain a very high local sector density in the centre of the region. A significant proportion of the shopping trips emanating from a suburb will be directed towards the regional centre, since customers here can benefit from low prices in the centre at relatively low transport costs. For zones which lie a long distance from the centre, virtually all shopping will take place within the zone. The local sector density will be high in the regional centre, low in suburbs and it will approach the regional average as the distance from the centre increases. Gjestland et al. (2006) find empirical support for such an intuitively appealing pattern from observations of Norwegian regions.

The level of local sector density at a centre reflects the central place system of the region and the importance and dominance of a centre can be argued to be a decreasing function of the average distance to potential customers outside the centre. Equations (3.11)–(3.14) show how this idea is implemented in the model.

The discussion so far means that the intra-regional distribution of (local sector) jobs reflects the residential location pattern. At the same time it makes sense to assume that residential location decisions are influenced by the job opportunities within a reasonable commuting distance. This means that the spatial distribution of jobs and people are interdependent. This is the fundamental mechanism in economic base modelling.

3.2.3 The Decision to Stay or Move from a Residential Site

The residential location choice can be considered to result from a two-step decision process. First, a household decides whether to move from the current residential site. Second, households moving have to choose between alternative locations. Consider first the diagonal elements of a matrix of transition, migration, probabilities.

One hypothesis incorporated into our model is that the probability of remaining in a zone is positively related to the labour market accessibility of the zone. This is consistent with the findings from Swedish micro-data (Lundholm, 2010; Eliasson et al., 2003), while Van Ham and Hooimeijer (2009) find a similar result for the Netherlands. The explanation is that labour market accessibility allows greater flexibility and can generally be seen as a favourable attribute for a residential location.

A main challenge in the model formulation then is to specify an operational measure of labour market accessibility. In this chapter accessibility is represented by a weighted-average distance to all other zones of the geography. Each zone is weighted by the number of jobs, adjusted for the competition for jobs, measured by the number of jobs as a proportion of the local number of job seekers. In addition, the weights involve a distance deterrence function that places a relatively high weight on destinations which lie within a short distance from the residential location. The measure is defined in equations (3.6)–(3.9). It will henceforth be referred to as 'average distance'.

Finally, the measure of average distance is combined with information on the local labour market situation in the function that determines the probability that workers move from a specific zone (equation 3.10). Assume that a zone has high unemployment. If this zone is centrally located in the region, with a low value of average distance, many workers will choose to commute rather than move from their current residential location. However, if the average distance is long, then migration will be a more frequent spatial interaction response to high unemployment.

3.2.4 Spatial Equilibrium and Migration Flows between Different Zones

The migration between different zones is modelled through the introduction of a search strategy where a worker evaluates destinations successively outwards over the network. The worker will move to the first place where the conditions are 'satisfactory'. Options further out in the network will then not be evaluated. Hence an absorption effect is introduced, analogously to the basic idea in the theory of intervening opportunities (Stouffer, 1940). This further means that the probability of moving decreases as the worker evaluates alternatives which lie progressively further out in the network.

Another central hypothesis within the regional science literature is that distance limits spatial interaction. Accounting for the absorption effect and the distance deterrence effect forms a symmetric matrix, which is next normalized into a migration probability matrix. This matrix is then used to find the equilibrium solution for the system. See Appendix section A5 for details.

3.2.5 The Relationship between the Spatial Distribution of Jobs and People, an Economic Base Multiplier Process

The spatial distribution of jobs is linked to the spatial distribution of people through labour market accessibility and the simultaneity between commuting flows and migration flows. The economic base mechanism represents the more direct link between the location of jobs and people. As mentioned above, location decisions of local sector firms reflect the shopping behaviour and the location pattern of the households demanding the goods and services being offered. At the same time workers employed in local sector firms tend to prefer a residential location close to the firm. Assume increased basic sector activity in a zone. This causes a rise in labour demand, attracts labour to the zone and increases the demand for goods and services produced in the local sector. This creates further demand for labour and initiates a positive growth cycle, known in the literature as an economic base multiplier process (equations 3.4–3.5).

The equilibrium modelling approach in this chapter accounts for different kinds of interdependencies in a simultaneous treatment of location decisions made by firms and households. This can be argued to be a preferred approach to introducing a specific causality on the employment–population interaction. According to Hoogstra et al. (2011), the nature of this causality differs across space and time. In a meta-analysis, they find that the empirical evidence is highly inconclusive on the jobs–people direction of causality, albeit most results point towards 'jobs follow people'.

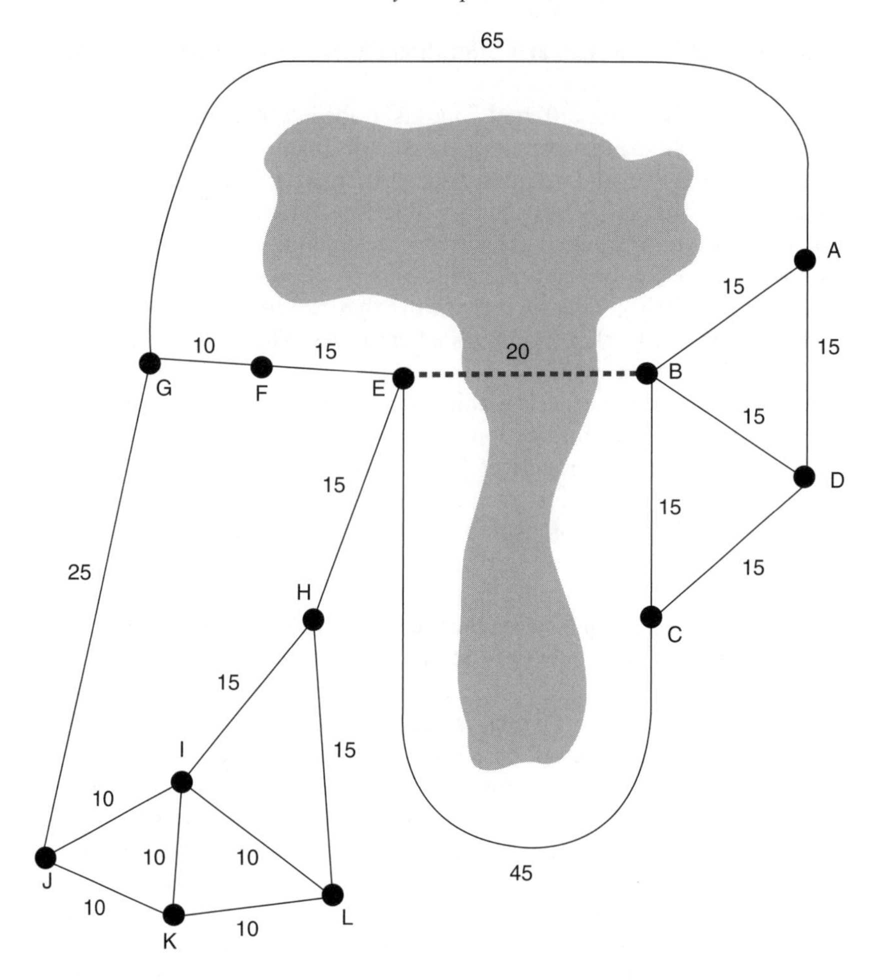

Figure 3.1 A network with 12 nodes and a topographical barrier; generalized distances of travelling on the different links are shown

3.3 THE GEOGRAPHY, EXPERIMENTS AND RESULTS

3.3.1 The Geography

In this chapter we utilize the geography presented in Thorsen et al. (1999, p. 355) and shown in Figure 3.1. The figures shown next to the links represent generalized distances. The generalized distance of travelling across the bridge represents the length of the bridge, the time taken to cross it and any other charges, for instance road tolls. The use of road tolls has some potentially important policy implications. McArthur et al. (2013) show that road tolls can deter partial interaction, while McArthur et al. (2012) consider the second-order impacts which may result from road tolls.

Only one mode of transport is considered as a simplifying assumption. It may not be entirely unreasonable for the rural and semi-urban geography which we consider, where public transportation systems are often lacking. The model could be expanded to account for additional modes of transport; however this is left for future research.

The region is comprised of 12 towns or zones and three clusters can be identified. The first consists of zones I, J, K, L and H. Zone I is defined as the central business district (CBD). This makes the other zones in this cluster suburbs. Zones A, B, C and D represent a rural area. Generalized costs incurred travelling from the peripheral area to the central area are high due to the presence of a lake. The third cluster of zones, E, F and G, are intermediate zones.

Our assumptions regarding the distribution of basic sector activity are shown Table 3.1. Initial values of population are also defined, although the resulting equilibrium is independent of these values. The total population of the region is 270 000. The equilibrium solution is presented in Table 3.1.

The distribution of the population and employment in this geography is as expected. Both population and employment are concentrated in the CBD and its surrounding zones. The population and employment in the most peripheral zones are lowest. In part this reflects the lower basic sector employment in these zones. However it also reflects the lower accessibility. Due to the distance of these rural zones from the rest of the zones in the geography, it is not feasible to live in these zones and commute to a workplace in the urban area. People therefore tend to live and work in the same zone, or commute to a nearby rural zone. This creates a somewhat isolated subsystem, which will be referred to as the rural area of the region.

Consider an attempt to decrease the degree of centralization within a region. If we consider either the average distance measure of accessibility used in this chapter, or some variant of the popular measure presented in Hansen (1959), then we see that it depends on two components. The accessibility of a particular zone will depend on the distribution of employment and the transportation infrastructure, technology and costs.

Policy-makers can potentially manipulate both employment and the transportation network. In this chapter, we consider the effect of changes to the transportation network. This may have the indirect effect of

Table 3.1 Number of basic sector jobs and the equilibrium population and employment

	Basic sector jobs	Population	Employment
A	1000	10935	8170
B	1000	11551	8488
C	1000	7775	5908
D	1000	11551	8488
E	7000	19044	15670
F	7000	19885	17907
G	7000	18918	16315
H	8000	22360	14499
I	20000	60207	120000
J	12000	28314	17986
K	12000	30538	18456
L	12000	28915	18113

redistributing employment. It is also possible to use the model to consider a situation where policy-makers can directly alter the distribution of employment. This is left for future research. In Norway, changes to the transportation network have been used extensively to improve accessibility levels in rural areas.

The infrastructure modelled in this chapter is a bridge linking zones B and E. This will give all of the rural zones a shorter route to the larger labour markets close to the CBD. It is natural to think of improved accessibility as a benefit. Indeed, there is evidence of this in the literature (Gjestland et al., 2012; Osland and Thorsen, 2008; Eliasson et al., 2003). An intuitive assumption would be that increasing the attractiveness of a zone would tend to increase its population and the level of employment. The answer is more complex, however, and has some important implications for spatial equity.

Calculating an accessibility measure, such as the potential measure given by Hansen (1959), shows that the accessibility of all zones is enhanced by the construction of the new infrastructure. This is in line with expectations, since improved connectivity allows easier access to employment opportunities located in other zones. It is worth noting however that a 'naïve' approach to predicting the change in accessibility will potentially induce an error. A naïve approach is defined here as assuming that the distribution of employment is exogenous to changes in the transportation network. If people and workplaces move in response to the new infrastructure, then the naive approach to predicting changes in accessibility will result in errors. The direction and the magnitude of these errors will depend on the spatial configuration of zones and the preferences of the zones' inhabitants.

3.3.2 Employment

Figure 3.2 shows the percentage change in population and employment in the four rural zones (A, B, C and D) following the installation of a road link between B and E. Different levels of generalized distance for this link are considered.

Figure 3.2 shows that, for a given set of parameters, the outcome for the rural zones varies dramatically in response to a change in the infrastructure, depending on the extent of the change. For a generalized distance of 60, the new infrastructure represents no change in accessibility and therefore no change to either population or employment. For a bridge with a generalized distance greater than 40, both employment and population decline slightly in response to the new bridge. This decline is in spite of the zones' increased accessibility.

Employment will, in fact, decline for any bridge with a generalized distance greater than 20. The decline is caused by a change in local sector activity. There are two mechanisms underlying this. The first is the decline in population. Worker migration is deterred by distance and by an absorption effect. The new bridge reduces both effects, resulting in some people

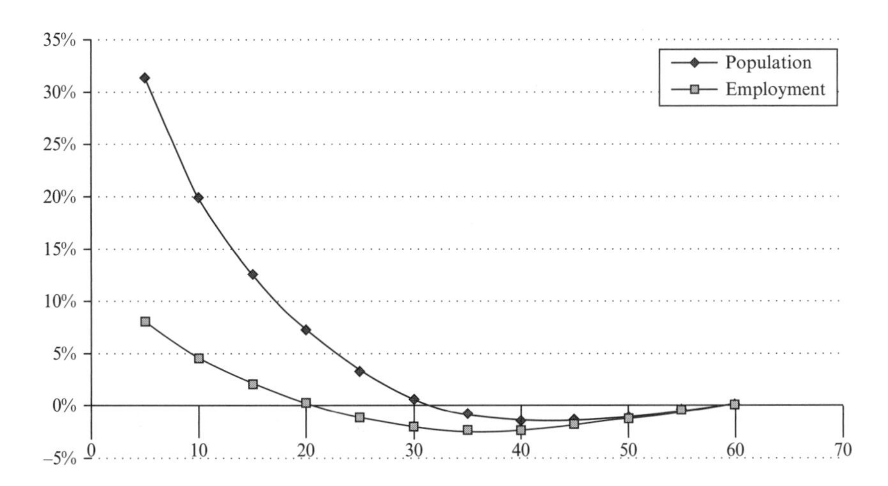

Figure 3.2 The percentage change in population and employment in the four rural zones (A, B, C and D) following the installation of a road link between B and E with different generalized distance

choosing to leave the rural area. As the population of the zones falls, there is less local demand to satisfy.

The second mechanism is that the bridge encourages shopping outside of the rural area, further decreasing local sector activity. The worst situation for employment occurs when a bridge with a generalized distance of 35 is opened. However, as the bridge becomes less expensive, commuting options improve and the rural area becomes more attractive as a residential location. This increases the population and hence the demand in the local sector. The first bridge which is unambiguously positive for the rural zones – that is, where total employment and population are increased – is one with a generalized distance of 20. Larger gains in both population and employment can be realized when the generalized distance is below 20.

This simple numerical experiment has shown that improving accessibility need not always result in an outcome which would be desirable from a regional policy perspective. In Norway, improving the accessibility of rural areas is a key way of achieving the policy objective of sustaining population and employment in the periphery. A large and expensive infrastructure project which resulted in a decline in both population and employment would not be considered successful when evaluated against this policy objective.

Figure 3.3 shows the change in employment in different groups of zones, for different bridges. Again, a bridge with a generalized distance of 60 is

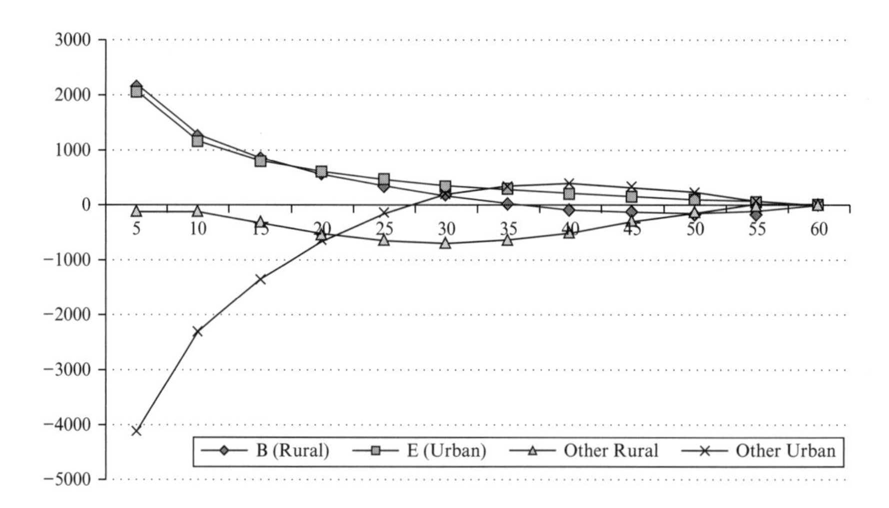

Figure 3.3 Change in employment in different groups of zones following the opening of a bridge with different levels of generalized distance

equivalent to not having a bridge at all. Four groups are defined. Two of the groups are comprised of single zones. These zones, B and E, are located at either side of the bridge. Ubøe and Thorsen (2002) highlight the important role played by such zones. They represent compromises between a desire to live in the rural area and a desire to access the urban labour market. The next of the two groups of zones is defined as the rural zones excluding B, that is A, C and D. The final group of zones includes all of the urban zones excluding E.

Consider the endpoints of the bridge, zones B and E. Zone E, which lies on the urban side of the new bridge, gains more employment than B when bridges with generalized distances greater than 20 are constructed. For these more expensive bridges, zone E offers easier access to the urban labour market than B, while still being relatively close to the rural zones. Both endpoints of the bridge benefit from all bridges which have a generalized distance of 35 or less. For a connection with a generalized distance of 5, these two zones can gain around 4000 new jobs.

From a regional policy perspective, it is important to consider the winners and losers in the situation. For a bridge with a generalized distance between 35 and 60, all the rural zones will lose out, as their workplaces move to the urban zones. For a generalized distance less than 35, both endpoints of the bridge will experience gains in employment. However, the other rural zones, A, C and D, lose employment for all lengths of bridge considered. The gains made by B and E therefore come at the expense of their rural neighbours. As stated previously, for there to be a net gain in employment, the bridge must have a generalized distance of 20 or less. This net gain in employment will occur as a result of B gaining employment, as it moves from the other rural zones. The employment loss for these other rural zones is maximized for a bridge with a generalized distance of 30. When the generalized distance is lower than this, employment is attracted away from the other urban zones.

So far, the numerical experiments have shown that installing a new bridge may have potentially undesirable consequences for employment, particularly with regard to its distribution. The distributional consequences in our model depend on the generalized distance of the new infrastructure. Bridges with the lowest generalized distance redistributed the most jobs away from the urban centre of the region towards the periphery. Bridges with an intermediate cost tended to favour the urban area and the most expensive bridges had a negligible effect.

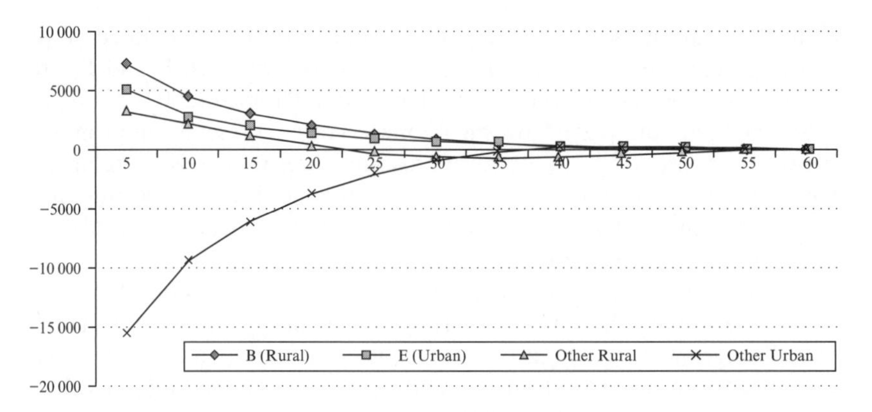

Figure 3.4 (*a*) *Urban/rural groupings*
Change in population in different groups of zones following the opening of a bridge with different levels of generalized distance

3.3.3 Population

In addition to considering the distribution of employment, it is also important to consider the distribution of the population. Figure 3.4(a) shows the population change in response to the construction of a bridge with different generalized distances for the same groupings of zones as in Figure 3.3. Unsurprisingly, the picture is rather similar. Bridges with a lower generalized distance tend to favour redistribution from the urban centre to the periphery, with bridges with a generalized distance greater than 35 having very minor effects on the redistribution of population. One important feature to emerge here is that in terms of population, it is possible for the other rural zones to gain. For bridges with a generalized distance less than 20, the rural area has a net gain of employment. From this same point, both zone B and the other rural zones experience gains in their population.

Considering the distribution of the population presents a slightly more positive outcome than when considering employment. If the increase in accessibility is sufficiently large, then it is possible to affect a rise in the population in the rural areas. The redistribution of employment away from the urban centre towards zones B and E put them within commuting distance of the other rural zones, making them more attractive residential locations.

Figure 3.4(b) shows the interesting case of zones B and D. Prior to the change in this infrastructure, these zones shared an equivalent position in the network. Both had an equal distance to all of the other zones in the network and hence had the same level of population prior to the new

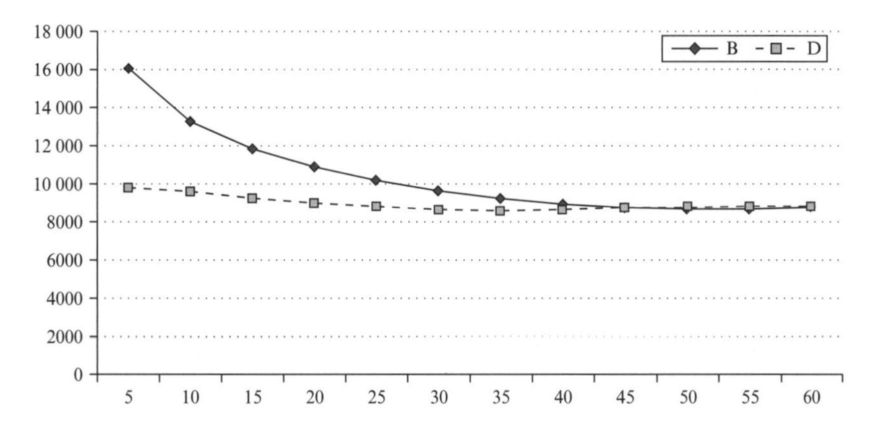

Figure 3.4 (b) The rural zones B and D

bridge. The new bridge causes potentially dramatic changes, depending on the generalized distance, with lower generalized differences causing bigger differences in population. For bridges with a generalized distance of over 35, changes in population are minor. The zones even lose some of their population for generalized distances above this level.

For shorter bridges, with generalized distances below 35, both zones B and D will experience a rise in population as a result of the new bridge. Zone B gains population at a faster rate than zone D. For a bridge with a generalized distance of 5, B will have a population increase of 82 per cent, while the population in D would increase by 12 per cent. The bridge has therefore taken two equivalent zones and caused them to become quite different. This highlights the pivotal role which can be played by small differences in relative accessibility.

3.3.4 Migration Sensitivity to Distance

The model has many parameters which can be changed to fit different contexts. The experiments presented in this chapter use standard values taken from various sources in the literature. During sensitivity analysis carried out, one parameter in particular proved to be important. This parameter is β, as shown in Appendix section A2, which is the distance deterrence parameter which appears in the migration function. A higher value increases the disutility experienced when moving. The equilibrium distribution of the population is the net result of a trade-off between this disutility and the utility gained from residing in a particular location. The standard value which has been used in this chapter was $\beta = -1$. Plane (1984) found a value of –1 for interstate migration in the US.

This parameter is important, since it plays an important role in determining how much relocation takes place in response to a change in the transportation infrastructure. This relocation refers both to population and employment. Sensitivity analysis, not presented here, showed that results are most sensitive to changes in β. The final effect of the bridge will depend on the value taken by this parameter. For example, when $\beta = 1$ and a new bridge with a generalized distance of 20 is opened, the rural zones will experience a population increase of 2400, with 57 extra jobs. When $\beta = 0.5$, the population will increase by 1741, with a loss of 666 jobs. When $\beta = 2.5$, the population increases by 2192, with 701 extra jobs.

The β parameter is therefore very important in determining the effects of the bridge. Note that this parameter affects both the before and the after situations. When the parameter is lower, the starting population in the rural area is higher. People are more willing to live on the periphery in this situation. When the parameter is higher generalized cost becomes more important and people become more sensitive to it. They therefore prefer not to live on the periphery and are more sensitive to changes in generalized cost. Therefore, a higher beta will give stronger relocation effects for a given infrastructure change.

Knowledge of the β parameter is therefore important in making an evaluation in any given context. It is also important to know whether this parameter takes a different value when considering intra-regional migration rather than interregional migration. The final results of any change in the infrastructure may vary quite radically with this parameter.

3.4 CONCLUSION

This chapter has proposed a model which facilitates the analysis of the effects of innovations in the transportation network. The model allows analysis of changing accessibility patterns, in terms of changes in both the transportation network itself, and the location of economic activity and workers. One of the key benefits of this model is that it allows the long-term implications of changes in the transportation network to be considered. For example, predicting changes in accessibility while assuming that the locations of workers and jobs is fixed may under- or overpredict the true accessibility effect.

A particularly important aspect explored is the effect of new infrastructure on the distribution of economic activity and people. Improving accessibility does not necessarily imply a more even distribution of either employment or population. Particular focus has been placed on the urban–rural dimension. Numerical experiments with the model have shown that while the accessibility of a rural area may be improved by a particular infrastructure project, the number of jobs located in the rural area may actually decline. Of course, the increased accessibility creates new opportunities for commuting, so the fall in the number of jobs need not imply a change in the zone's employment rate. Increased accessibility may change the role of a zone from a rural village to a suburb, where the population exceeds the number of jobs.

A number of important conclusions emerge therefore, both for analysts and for policy-makers. Analysts must be cautious of taking a naive approach to accessibility, where locations are assumed to be fixed. Many infrastructure projects may be undertaken with an explicit aim of inducing relocations. The accessibility modelling framework used should reflect this. The role of time should also be taken into account. While in the short term locations may be relatively fixed, in the long run we may expect different effects. Advice given to policy-makers based on accessibility analysis should distinguish between the short- and long-term anticipated effects. Such dynamic analysis is possible with the model presented in this chapter.

Policy-makers must be clear about what they want to achieve with new infrastructure projects. Increasing the accessibility of a region, particularly a peripheral region, is often seen as unambiguously good. While this may be true for the wider region, for specific zones this may not be the case. The experiments in this chapter have shown that accessibility improvements may be accompanied by falling population and employment. In another case, the model showed that some zones experienced falling employment but rising population. Policy-makers must be clear about what they would consider successful outcomes from their transport investment, as well as having realistic expectations about what is possible.

In the cases considered in this chapter, the rural areas were generally better off for bridges with the lowest generalized distance, which provided the biggest increase in accessibility. The generalized distance will include some factors which are difficult for policy-makers to control, the biggest factor being distance. However, the travel time may be one policy variable. New infrastructure should be suitable for the volume of traffic expected so that congestion does not increase generalized costs. Another important variable is road pricing. In some countries, such as Norway, new infrastructure is often built to improve the accessibility of rural areas. However, these projects are usually financed by relatively high road tolls. This increased cost deters spatial interaction and reduces the potential accessibility gains. This could mean the difference between a favourable and unfavourable outcome for the rural zones. In such a case, the road tolls directly contradict the policy aim of improving accessibility and spatial equity.

Many analyses of accessibility use measures based on potential commuting. This implicitly assumes that workers will fix their location and respond to improved accessibility through commuting. This assumption may be correct in some instances, but may also under- or overestimate the true effect on accessibility of a change in the transportation network. It may be helpful here to distinguish between what we might call the pure network accessibility effect and the overall or net accessibility effect.

The network accessibility here refers to the increase in accessibility resulting from a change in the transportation network, *ceteris paribus*. If we increase the network accessibility, this may lead to a new location pattern and hence a change in the weights given to the nodes in the network. If this new location pattern favours the areas with an improved network accessibility, then the total accessibility of this area will be further enhanced. If, as in many of our simulations, jobs and workers migrate out of the zone with improved network accessibility, it may experience a decrease in overall accessibility.

In our presentation of the model, we have focused on how population and employment change in different zones in response to a change in the transportation infrastructure. Underpinning this has been an assumption that workers and jobs are homogeneous. However, the model is capable of dealing with heterogeneous workers and jobs. In such an implementation of the model, changes in the transportation infrastructure may have different effects not only on different regions, but also on different groups of workers. These groups may have different preferences and different sensitivities to distance. For instance, a particular project may benefit high-income workers most, while another may favour low-income workers.

It is important to note one of the omissions from our model. We do not account for the housing market. We can expect that accessibility capitalizes into house prices. Higher house prices, or a lack of supply, might be cited as a reason why people do not relocate to accessible areas. However, in our model this effect is not present. Still, we see that it is not certain that people will want to move to an area where the accessibility is improved. It is also important to note that different factors may play different roles at different spatial scales. For example, on the intra-regional scale that we consider, the relocation of local sector activity plays an important role. At an interregional scale, this effect may prove to be much weaker. Such complex interactions should be considered in future accessibility analyses.

REFERENCES

de Bok, M. (2009), 'Estimation and validation of a microscopic model for spatial economic effects of transport infrastructure', *Transportation Research Part A: Policy and Practice*, **43** (1), 44–59.

Eliasson, K., U. Lindgren and O. Westerlund (2003), 'Geographical labour mobility: migration or commuting?', *Regional Studies*, **37** (8), 827–837.

Fujita, M., P. Krugman and A. Venables (1999), *The Spatial Economy: Cities, Regions and International Trade*, Cambridge, MA: MIT Press.

Geurs, K. and B. Van Wee (2004), 'Accessibility evaluation of land-use and transport strategies: review and research directions', *Journal of Transport Geography*, **12** (2), 127–140.

Gjestland, A., D. McArthur, L. Osland and I. Thorsen (2012), 'A bridge over troubled waters: valuing accessibility effects of a new bridge', in K. Geurs, K. Krizek and A. Reggiani (eds), *Accessibility Analysis and Transport Planning: Challenges for Europe and North America*, Cheltenham, UK and Northampton, MA, USA: Edward Elgar, pp. 173–193.

Gjestland, A., I. Thorsen and J. Ubøe (2006), 'Some aspects of the intraregional spatial distribution of local sector activities', *Annals of Regional Science*, **40** (3), 559–582.

Glenn, P., I. Thorsen and J. Ubøe (2004), 'Wage payoffs and distance deterrence in the journey to work', *Transportation Research Part B: Methodological*, **38** (9), 853–867.

Goldner, W. (1971), 'The Lowry model heritage', *Journal of the American Institute of Planners*, **37** (2), 100–110.

Hansen, W. (1959), 'How accessibility shapes land use', *Journal of the American Institute of Planners*, **25** (2), 73–76.

Holl, A. (2007), 'Twenty years of accessibility improvements: the case of the Spanish motorway building programme', *Journal of Transport Geography*, **15** (4), 286–297.

Hoogstra, G., J. van Dijk and R. Florax (2011), 'Determinants of variation in population–employment interaction findings: a quasi-experimental meta-analysis', *Geographical Analysis*, **43** (1), 14–37.

Hou, Q. and S. Li (2011), 'Transport infrastructure development and changing spatial accessibility in the Greater Pearl River Delta, China, 1990–2020', *Journal of Transport Geography*, **19** (6), 1350–1360.

Krugman, P. (1991), 'Increasing returns and economic geography', *Journal of Political Economy*, **99** (3), 483–499.

Liu, S. and X. Zhu (2004), 'An integrated GIS approach to accessibility analysis', *Transactions in GIS*, **8** (1), 45–62.

López, E., J. Gutiérrez and G. Gómez (2008), 'Measuring regional cohesion effects of large-scale transport infrastructure investments: an accessibility approach', *European Planning Studies*, **16** (2), 277–301.

Lowry, I. (1964), *A Model of Metropolis*, Santa Monica, CA: RAND Corporation.

Lundholm, E. (2010), 'Interregional migration propensity and labour market size in Sweden, 1970–2001', *Regional Studies*, **44** (4), 455–464.

Mackiewicz, A. and W. Ratajczak (1996), 'Towards a new definition of topological accessibility', *Transportation Research Part B: Methodological*, **30** (1), 47–79.

Martín, J. and A. Reggiani (2007), 'Recent methodological developments to measure spatial interaction: synthetic accessibility indices applied to high-speed train investments', *Transport Reviews*, **27** (5), 551–571.

McArthur, D.P., G. Kleppe, I. Thorsen and J. Ubøe (2013), 'The impact of monetary costs on commuting flows', *Papers in Regional Science*, **92** (1), 69–86.

McArthur, D., I. Thorsen and J. Ubøe (2012), 'Labour market effects in assessing the costs and benefits of road pricing', *Transportation Research Part A: Policy and Practice*, **46** (2), 310–321.

Meijers, E., J. Hoekstra, M. Leijten, E. Louw and M. Spaans (2012), 'Connecting the periphery: distributive effects of new infrastructure', *Journal of Transport Geography*, **22**, 187–198.

Nævdal, G., I. Thorsen and J. Ubøe (1996), 'Modeling spatial structures through equilibrium states for transition matrices', *Journal of Regional Science*, **36** (2), 171–196.

Osland, L. and I. Thorsen (2008), 'Effects on housing prices of urban attraction and labor-market accessibility', *Environment and Planning A*, **40** (10), 2490–2509.

Plane, D. (1984), 'Migration space: doubly constrained gravity model mapping of relative interstate separation', *Annals of the Association of American Geographers*, **74**, 2.

Reggiani, A., P. Bucci and G. Russo (2011), 'Accessibility and impedance forms: empirical applications to the German commuting network', *International Regional Science Review*, **34** (2), 230.

Stouffer, S. (1940), 'Intervening opportunities: a theory relating mobility and distance', *American Sociological Review*, **5** (6), 845–867.

Straatemeier, T. (2008), 'How to plan for regional accessibility', *Transport Policy*, **15** (2), 127–137.

Thorsen, I., J. Ubøe and G. Nævdal (1999), 'A network approach to commuting', *Journal of Regional Science*, **39** (1), 73–101.

Ubøe, J. and I. Thorsen (2002), 'Modelling residential location choice in an area with spatial barriers', *Annals of Regional Science*, **36** (4), 613–644.

Van Ham, M. and P. Hooimeijer (2009), 'Regional differences in spatial flexibility: long commutes and job related migration intentions in the Netherlands', *Applied Spatial Analysis and Policy*, **2** (2), 129–146.

APPENDIX: A TECHNICAL PRESENTATION OF THE MODEL

This Appendix provides a technical presentation of the mechanisms presented in section 3.2. The spatial distribution of basic sector firms is considered to be exogenously given. In this version of the model we further ignore the possibility that migration decisions are affected by job diversity and local amenities, while housing prices and wages are assumed to be exogenously given.

A1 Basic and Local Sector Firms: The Economic Base Multiplier

Total employment in zone i (E_i) is defined as the sum of basic sector employment (E_i^b) and local sector employment (E_i^l) in the zone:

$$E_i \equiv E_i^b + E_i^l \tag{3.1}$$

Let **L** be a vector representing a given residential location pattern of workers, while T_{ij} is the probability that a worker lives in zone i and works in zone j. Hence $T = [T_{ij}]$ represents the commuting matrix in the geography and by definition:

$$\boldsymbol{TE} = \boldsymbol{L} \tag{3.2}$$

The spatial distribution of local sector activities reflects both the spatial residential pattern and the spatial shopping behaviour. Assume that the number of workers living in a zone is proportional to the number of residents/consumers in the zone, and let C_{ij} be the number of local sector jobs in zone i which are supported by shopping from workers living in zone j. Hence $\boldsymbol{C} = [Cij]$ is a shopping matrix and the spatial distribution of employment in local sector activities is given by:

$$\boldsymbol{E^l} = \boldsymbol{CL} \tag{3.3}$$

Given that the inverse of the matrix $(\boldsymbol{I} - \boldsymbol{TC})$ exists, it follows from equations (3.1), (3.2) and (3.3) that:

$$\boldsymbol{L} = (\boldsymbol{I} - \boldsymbol{TC})^{-1}\boldsymbol{TE^b} \tag{3.4}$$

$$\boldsymbol{E^l} = \boldsymbol{C}(\boldsymbol{I} - \boldsymbol{TC})^{-1}\,\boldsymbol{TE^b} \tag{3.5}$$

These solutions capture the economic base multiplier process: people attract local sector activities, while local sector employment opportunities attract workers. The spatial distribution of basic sector activities ($\boldsymbol{E^b}$) will be considered exogenous in the model, while the other variables ($\boldsymbol{C}$, $\boldsymbol{E^l}$, $\boldsymbol{T}$, $\boldsymbol{L}$) are represented by a set of equations representing shopping, commuting, location and migration decisions of households and firms.

A2 Interzonal Migration Flows and Spatial Equilibrium

In modelling migration probabilities, Nævdal et al. (1996) introduced a nice trick to facilitate the construction of Markov chains. The construction uses a symmetric matrix $\boldsymbol{Q} = \{Q\}_{i,j=1}^{N}$, where all the elements are dependent on the characteristics of the geography. Nævdal et al. (1996) showed that any assumption about the coefficients Q_{ij} can be interpreted as an assumption about migration flows in the equilibrium state.

As a next step Nævdal et al. (1996) introduced some network characteristics which are symmetric between zones and which are relevant in explaining the relevant kind of spatial interaction. For a connected network with fixed Q_{ij}-s, the construction produces regular Markov chains and these chains always have unique equilibria. We have been using the n-step model as specified in Nævdal et al. (1996). See that paper for more detail about the construction.

In our chapter the coefficients are state-dependent, that is, the transition probabilities are functions of **E** and **L**. In that case the equilibria are no longer unique, but the interpretation in terms of the strength of migration flows in the equilibrium state remains valid. See Nævdal et al. (1996).

A3 The Decision to Stay or Move from a Zone

It is a central hypothesis in the model that the decision to stay or move from a zone depends on the labour market accessibility of the zone. Labour market accessibility is introduced by a measure of average distance. The average distance from zone i is given by:

$$d_i = \sum_j \frac{W_j}{\sum_k W_k} d_{ij} \tag{3.6}$$

Labour market accessibility is not just a matter of distances; the weights W_i represent the size of alternative destinations. The size and thickness of a potential destination are assumed to be represented by the number of jobs; $W_j = E_j$, $j = 1, 2, \ldots, N$, defining d_i as the average distance to potential employment opportunities in the geography. In a spatial labour market context, however, it can be argued that potential destinations within a reasonable commuting distance should be given more weight than more distant destination alternatives. This is done through the introduction of a distance deterrence function (d_{ij}), that places a relatively high weight on destinations which lie within a short distance from the residential location:

$$W_j = E_j(1 - D(d_{ij})) \tag{3.7}$$

The distance deterrence function and weights are parameterized by d_∞, d_0 and μ in the following logistic expression:

$$D(x) = \frac{1}{1 + e^{-k(x - x_0)}},\ x_0 = \frac{1}{2}(d_0 - d_\infty),\ k = \frac{2\log\left(\frac{1}{\mu} - 1\right)}{d_\infty - d_0} \tag{3.8}$$

d_∞ is the upper limit for how far workers, as a rule, are willing to commute on a daily basis, d_0 is the lower limit (internal distance) where people are insensitive to further decreases in distance, while μ captures friction effects in the system. The values of x_0 and k are given to satisfy the conditions $D(d_0) = \mu$ and $(1 - D(d_\infty)) = \mu$. If, for example, $\mu = 0.05$, this means that the function will fall to 5 per cent of its value outside the range where $d_0 \leq x \leq d_\infty$. Glenn et al. (2004) give a microeconomic and geometric justification for the use of such a function.

Finally, the definition of average distance also accounts for the competition for jobs at alternative locations (Liu and Zhu, 2004), represented in the model by the proportion of the total number of job seekers in each potential destination, $\frac{E_j}{L_j}$:

$$W_j = E_j(1 - D(d_{ij}))\frac{E_j}{L_j} \tag{3.9}$$

The definition of average distance is included in the diagonal elements of the migration matrix, reflecting workers' spatial interaction response to an unfortunate local labour market situation ($L_i > E_i$). A high value of d_i (and $D(d_i)$) means that the migration decisions are very sensitive to the local labour market situation. On the other hand, high local unemployment does not in itself bring about a significant outmigration from zones in highly accessible labour market locations (low d_i), with an excellent commuting potential. This is captured by the following specification of α_i:

$$\alpha_i = \alpha_i(L_i) + D(d_i)\max\left\{\rho\left(\frac{L_i - E_i}{L_i}\right), 0\right\} \tag{3.10}$$

Here, the parameter ρ reflects the speed of adjustment to an unfortunate labour market situation, towards a situation with a balance in the local labour market, $L_i = E_i$.

A4 The Spatial Distribution of Local Sector Employment

It is reasonable to assume that local sector activities in a whole region (E_r^l) are proportional to population in the region (L_r):

$$E_r^l = \sum_i^n E_i^l = b\sum_i^n L_i = bL_r b > 0 \tag{3.11}$$

where b is the proportion parameter. Let the spatial distribution of local sector employment be represented by $\frac{E_i^l}{L_i}$, that is the number of shop employees per resident at location i. Assume, as a simplification, a monocentric region offering agglomeration benefits for local sector firms and price savings for households in shopping. Shopping decisions then result from a trade-off between price savings and transport costs.

Transportation costs provide an incentive for local sector firms to decentralize in order to cater for local demand. The trade-off between transport costs and potential price savings plays a central role in Gjestland et al. (2006), providing a theoretical base in favour of the hypothesis that the frequency of shopping locally is a smooth, concave function of the Euclidean distance from the CBD.

In our chapter we assume that there is only one CBD and define the local sector density by:

$$\textit{Local sector density} = \frac{E^l}{L}(\textit{distance to CBD}) \tag{3.12}$$

$$= R_\infty(1-\exp[-\beta_{cbd}\cdot \textit{distance to CBD}])+C\cdot\exp\left[-\left(\frac{\gamma\cdot \textit{distance to CBD}}{d_{dispersion}}\right)^2\right]$$

The only free parameter is β_{cbd} which controls the decay in the local sector density curve. The other parameters are defined as follows:

$$R_\infty = \frac{\sum_{i=1}^N E_i^l}{\sum_{i=1}^N L_i} \; (\textit{Average local sector density in the system as a whole}) \tag{3.13}$$

$d_{dispersion}$ is the spatial extension of the CBD, $\gamma = \sqrt{-\ln[0.05]}$ forces the effect of the second term of (12) down to 5 per cent of its peak value at the boundary of the CBD. Given values for β_{cbd}, R_∞ and $d_{dispersion}$, C is chosen such that the integral:

$$\int_0^{d_{dispersion}} Local\ sector\ density(r) \cdot \frac{2r}{d^2_{dispersion}} \cdot L_{cbd}\,dr = E^l_{cbd} \qquad (3.14)$$

The spatial distribution of local sector activities reflects the net effect of the price savings resulting from agglomeration forces and the transport costs of shopping in the CBD rather than locally.

A5 An Iterative Process towards Spatial Equilibrium

We begin with more or less random initial values for employment and population ($\boldsymbol{E}_0 = \boldsymbol{E}^l_0 + \boldsymbol{E}^b_0$ and $\boldsymbol{L}_0$). These values are fed into a state-dependent migration matrix **M** and adjusted to fit a local sector density curve. This is then iterated until we find a fixed point **L**, which represents the equilibrium solution for population (workers), that is, that **ML** = **L** and equilibrium values fitting the local sector (jobs) $\boldsymbol{E}^l$ to the local sector density curve.

4. High resolution accessibility computations

Thomas W. Nicolai and Kai Nagel

4.1 INTRODUCTION

Researchers assert that accessibility has a measurable impact in the real world: Hansen (1959) shows that areas which have more access to opportunities have a greater growth potential in residential development; Moeckel (2006) asserts that the principal idea of Hansen's approach is also true for businesses. In other words, locations with easier access to other locations are more attractive compared to otherwise similar locations with less access.

Many quantitative indicators can be used for accessibility (e.g. Geurs and Ritsema van Eck, 2001; Geurs and van Wee, 2004). Some examples are the distance to the next shopping area, travel time (e.g. Vandenbulcke et al., 2009) or distance (Borzacchiello et al., 2010) to the next railway station, or the number of opportunities (for example, workplaces, places to shop) within, say, 1 kilometre. Accessibility can be seen as the result of the following four components (Geurs and Ritsema van Eck, 2001; Geurs and van Wee, 2004):

1. A land-use component that deals with the number and spatial distribution of opportunities.
2. A transport component, which describes the effort to travel from a given origin to a given destination.
3. A temporal component, which considers the availability of activities at different times of day, for example in the morning peak hours.
4. An individual component that addresses the different needs and opportunities of different socio-economic groups, for example different income groups.

Accordingly, accessibility measures can concentrate on one or several of these components (Geurs and Ritsema van Eck, 2001):

1. The infrastructure-based approach is based on the performance of the transport system. An example would be the average speed by mode at certain locations.
2. The activity-based measurement deals with the distribution of possible activity locations in space and time. An example would be the number of shopping locations or workplaces within a certain distance. Alternatively, we could look at the number of shopping locations or workplaces within a certain travel time, which would combine the infrastructure-based with the activity-based approach.
3. A utility-based measurement of accessibility reflects the (economic) benefits, or maximum expected utility, that someone gains from access to spatially distributed opportunities (Geurs and Ritsema van Eck, 2001; de Jong et al., 2007). The typical example is the logsum term, discussed below.

Normally, accessibilities are attached to locations i. We can say that 'location i is very accessible', that is, it is easy to reach, presumably from many different locations. Alternatively, we can say that 'location i has good accessibility to certain services', meaning that starting at location i, we are able to get to many destinations. The present chapter will concentrate on the second variant, 'outgoing' accessibility; the question of how far this is interchangeable with 'incoming' accessibility will be addressed towards the end.

Furthermore, it will be assumed that our (quantitative) accessibility measure is of the mathematical form:

$$A_i = g(\sum_j a_j f(c_{ij})), \tag{4.1}$$

where the sum goes over all possible destinations (opportunities) j, a_j is an indicator of the attractiveness of the opportunity, c_{ij} is the generalized cost of travel to get from i to j, $f(c)$ is an impedance function that typically decreases with increasing distance and $g(.)$ is an arbitrary, but typically monotonically increasing, function. The accessibility at i is thus computed from a weighted sum over all possible destinations, where the weight is the product of the destination's attractiveness and how easy it is to get there. This form subsumes most if not all of the measures discussed above.[1] It includes the land-use component, since a_j is larger than zero only at locations where the opportunity exists and it includes the transport component via c_{ij}. It also includes the temporal component, since for practical computations the generalized cost of travel depends on the time of day, and it could depend on the individuals by making attractiveness and impedance dependent on their demographic attributes. A collection of different quantitative accessibility measures is, for example, given by Martín and Reggiani (2007, Table 4.1).

Often, both the origin location i and the destination locations j are assumed to be relatively large zones. On the destination side this is typically achieved by aggregating over all opportunities at zone z, obtaining:

$$A_i = g(\sum_j a_j f(c_{ij})) \approx g(\sum_z \sum_{j \in z} a_j f(c_{iz})) = g(\sum_z a_z f(c_{iz})), \quad (4.3)$$

where c_{iz} now is the generalized cost to travel from location i to zone z, and a_z is the aggregated attractiveness indicator of zone z. The aggregation issue that c_{iz} is not exactly the same as c_{ij} could be noted here; this can cause problems for origins close to the zone, see below.

There are also issues on the origin side when using zones. For example, one could start the computation from a 'typical' point in the zone, for example the zone centroid. Alternatively, one could start from all possible locations inside the zone and then average. The latter is prohibitively expensive to compute, as will become clear from this chapter. However, using just one point as typical for a zone is quite problematic since it is already intuitively clear that shifting such a point by, say, 500 metres can change accessibility by walking quite significantly.

Finally there is the issue that the aggregation of the destinations is problematic in the vicinity of the origin. For example, assume that you are in a zone without any opportunities, but close to a zone with many opportunities. Here it makes no sense to assume that one has to travel to the centroid of the neighbouring zone where then one finds all opportunities; some opportunities may in fact be much closer. In the same vein, there has always been the issue of how to deal with opportunities that are in the same zone as the origin, sometimes called the issue of self-potential (e.g. Frost and Spence, 1995; Fröhlich and Axhausen, 2004; Condeço-Melhorado et al., 2011): should one assume that these opportunities have distance zero to the origin, that is, that $c_{ij} = 0$ for those destinations? Or should one assume a typical distance/travel time/generalized cost of travel depending on the average zone size? Or different values for every zone? All of these somewhat arbitrary choices lead to different quantitative results, and because opportunities close to the origin have high weight the differences can be quite important. In the present chapter it will be shown that such choices do not need to be made and instead accessibility can be computed with high enough resolution so that these issues do not arise.

In consequence, the present chapter will, except for comparison purposes, assume that accessibility values are attached to points (see also Kwan, 1998) rather than zones. The accessibility computation for location *i* will thus include, for each opportunity *j*, a generalized cost to reach the network, a generalized cost to travel on the network and a generalized cost from the network to the opportunity. The network travel times will be based on the output of a dynamic traffic assignment, thus including congestion effects. This is similar in spirit to Kwan (1998), but goes beyond that contribution by: (1) including the generalized cost of network access and egress; (2) using a much higher spatial resolution; and (3) using congested travel times from a dynamic assignment.

In fact, our whole perspective comes from simulation where we assign activity-based demand patterns dynamically onto the network (Balmer et al., 2009). Thus the framework which evolves such patterns over many iterations, running a microscopic simulation of travel behaviour in each iteration, is already there. The challenge is to generalize those behavioural patterns, for example to obtain the generalized cost of travel to opportunities that none of the persons starting at location *i* ever visited. A second challenge comes from the SustainCity project (http://www.sustaincity.org), within which the present investigation was performed. The project's travel model was expected to return accessibility values at parcel level. This will be discussed in somewhat more detail below.

Once A_i is accepted as a measure relating to a single point, one can take this argument further and consider accessibility as a field, that is, continuously varying in space, $A(x, y)$, where x and y are the coordinates. As is common in many areas of science such fields can be visualized by calculating the values on regular grid points and then using an averaging plotting routine. Related approaches by other researchers fall into two classes:

1. In the first class, accessibility is treated as a continuous field, but the accessibility measure is not computed on the congested network. For example, Borzacchiello et al. (2010) use Euclidean distance. Bono and Gutiérrez (2011) do not use an accessibility measure similar to the ones discussed here, but measure the difficulty to reach the network.
2. In the second class, accessibility is not treated as a continuous field; instead, the nodes of the transport network are used as origins. In return, many of these computations use the network, albeit often not a network with time-dependent congestion.

Many of these (e.g. Gutiérrez and Gómez, 1999; Gutiérrez et al., 2010; Condeço-Melhorado et al., 2011) then interpolate between these nodes. However, it is clear that at the microscopic level one can improve on such an approach: accessibility between nodes should be lower than at nodes and not in between the two values. Also, sometimes it is not fully clear which interpolation scheme was used.

Arguably most similar to our approach are Liu and Zhu (2004) and Zhu and Liu (2004). The main difference is that their approach is embedded into a geographic information system (GIS) tool whereas ours is attached to an activity-based dynamic traffic assignment. This makes their version (presumably) easier to use, while ours (presumably) has the more realistic time-dependent congestion patterns. Also, their origins and destinations are zones or zone centroids, which are subject to the issues discussed earlier. Finally their papers do not present computing times.

The goals of this chapter are accordingly:

- to define an accessibility measure for each point of the region;
- to clarify how access to and egress from the transport network are handled;
- for grid-based computations, to clarify the effect of different spatial resolutions;
- to consider computational performance issues;
- to perform illustrative sensitivity studies in a real-world setting including the effect of traffic congestion.

Section 4.2 describes in detail how this accessibility measurement is implemented. Section 4.3 describes a real-world scenario to which the approach was applied. Section 4.4 presents results of this application, in particular with respect to spatial resolution and computing times. The chapter ends with a discussion and conclusions in sections 4.5 and 4.6.

4.2 METHODOLOGY: HIGH RESOLUTION ACCESSIBILITY

4.2.1 Accessibility Indicator

This chapter will specifically look at accessibility indicators of the form:

$$A_i := \ln \sum_k e^{V_{ik}}, \tag{4.4}$$

where k goes over all possible destinations and V_{ik} is the disutility of travel in order to get from location i to location k. Compared to equation (4.1), this thus uses $g(.) = ln(.)$, $a_j = 1$, $f(c_{ij}) = e^{-c_{ij}}$ and $-c_{ij} = V_{ij}$. This so-called logsum term has an econometric interpretation as the expected maximum utility (e.g. Ben-Akiva and Lerman, 1985).

The logsum term includes a land-use component that considers the number and distribution of opportunities and a transport component that determines the effort to get there. It can also be interpreted as (the logarithm) of a weighted sum over opportunities, where each opportunity is weighted with *eVik*, that is, by how easy it is to get there. Clearly in this formulation no other attributes of the opportunities are included.

It may be useful to recall the origins of equation (4.4). For this, assume that the full utility of location k, seen from i, is $U_{ik} = V_{base} + V_{ik} + \varepsilon_{ik}$, where V_{base} is a constant base utility for doing the activity at any location, V_{ik} is the systematic (= observed) disutility to get there and ε_{ik} is a random term which absorbs the randomness of the travel disutility, but more importantly the utility fluctuations around V_{base} of doing the activity. Under the typical assumption that the ε_{ik} are independent and identically Gumbel-distributed random variables, the expectation value of U_{ik} becomes:

$$E(U_i) = E(\max_k U_{ik}) = Const + \ln \sum_k e^{V_{ik}} \equiv Const + A_i. \quad (4.5)$$

Const is an integration constant which can, in principle, be computed. It contains both the effect of the base utility, V_{base} and some constants related to the Gumbel distribution. Since it is the same for all locations, it is typically dropped. As a result A_i can become negative.

Equation (4.4) sometimes includes a so-called scale parameter. See section 4.5 for a discussion of this.

It is important to note that the sum will go over all opportunities k separately. For example, if there are N_l opportunities at a location, the term $e^{V_{il}}$ will be added N_l times. Clearly this leads to the same result as:

$$\ln \sum_l N_l e^{V_{il}}, \quad (4.6)$$

that is, multiple opportunities at the same location can be aggregated right away. The following investigation will instead use a method that also aggregates over close-by locations, reducing computing time even more. For the time being, it is best to assume that each opportunity is included separately.

Accessibilities in the present chapter are computed based on a congested road network with time dependent travel times. This task is part of an attempt to couple a land use model, UrbanSim (Waddell, 2002; Miller et al., 2005; OPUS User Guide, 2011), with a transport model, MATSim (Balmer et al., 2005; Raney and Nagel, 2006; Balmer et al., 2009). In that configuration MATSim performs a traffic flow simulation based on the land-use and commuting patterns provided by UrbanSim. The inputs from UrbanSim consist of a list of persons, with references to households and, if applicable, to jobs. Both the households and the jobs refer, via intermediate steps, to locations, that is, to zones or to parcels. Additionally, road network information is read by MATSim. The current version of MATSim4UrbanSim constructs simple home–work–home commuting plans from that information, which are then used as the initial demand for MATSim. MATSim relaxes that demand over typically 100 iterations, performing both a route and a departure time assignment. As a result MATSim possesses a congested road network with time dependent travel times. The feedback to UrbanSim currently consists of the following indicators: (1) zone-to-zone travel times and generalized travel costs; (2) accessibility (as in the present chapter) for each zone or each parcel, where parcels correspond to individual lots; and (3) each simulated person's travel time and generalized travel cost of the commute. A comprehensive description of the simulation and integration approach of MATSim and UrbanSim is given by Nicolai and Nagel (2012, p. 21).

4.2.2 Accessibility of Location *i*

Overview

In order to calculate the accessibility A_i, origin location *i* and opportunity locations *k* are assigned to a congested road network with time-dependent travel times. For every given origin *i* a so-called 'least cost path tree' computation runs through the network and determines the best route and thus the least negative travel utility V_{ik}, to each opportunity location *k* by using the Dijkstra shortest path algorithm (Dijkstra, 1959). The best route from *i* to *k* depends on the given generalized cost such as link travel times or distances. Once the least cost path tree has explored all nodes, the resulting disutilities V_{ik} for all opportunities are queried and the accessibility is calculated as stated in equation (4.4).

Assignment of locations to the network

Origin and opportunity locations do not necessarily lie on the network. Thus the calculation of V_{ik} includes the disutility of travel to overcome the gap between locations and the road network.

For origin locations *i*, the distance to the network is given by either: (i) the Euclidean distance to the nearest node; or (2) the orthogonal distance to the nearest link on the network. If the mapping of location *i* is to a link, as in case (2), V_{ik} further includes the travel disutility to overcome the distance to the nearest node. The travel costs on the link are calculated by dividing the distance to the node by the travel speed of the considered transport mode, for example car (free speed or congested car travel times at a given time of day), bicycle or walking.

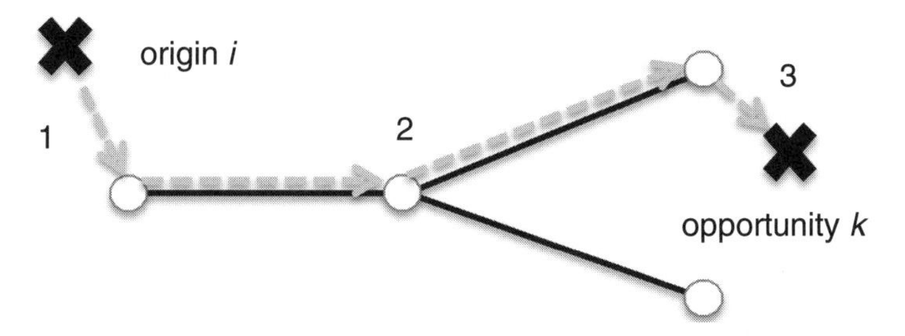

Figure 4.1 *The composition of the travel disutility v_{ik} consists of three parts: (1) the disutility to reach the network from i; (2) the disutility on the network; and (3) the disutility to reach opportunity k from the network.*

For opportunity (= destination) locations k, the Euclidean distance to the nearest node is used to determine the distance to the network.

Disutility of travel

As stated in equation (4.4), the computation of the accessibility for a given origin location i contains a summation of the term $e^{V_{ik}}$ for all opportunity locations k. The determination of the disutility of travel, V_{ik}, consists of the following contributions, as depicted in Figure 4.1:

$$V_{ik} := \beta_{tt_{wlk}} \cdot tt_{wlk,gap,i} + \beta_{tt_{mode}} \cdot tt_{mode} + \beta_{tt_{wlk}} \cdot tt_{wlk,gap,k}, \tag{4.7}$$

where:

tt_{wlk} is the travel time on foot.

tt_{mode} is the travel time according to the given transport mode. In the current implementation, transport modes are either car (free speed or congested), public transit, bicycle or walking.

$tt_{wlk,\ gap,\ i}$ is the travel time on foot to overcome the gap between the origin location i and the road network.

$tt_{wlk,\ gap,\ k}$ is the travel time on foot to overcome the gap between the road network and the opportunity location k.

β_{ttmode} and β_{ttwlk} are marginal utilities that convert travel times into utils. By default all marginal utilities are set to −12 utils/h. In MATSim terms, this is the sum of the marginal opportunity cost of time (typically −6 utils/h) and the marginal additional disutility of travel (typically another −6 utils/h).

The congested car speed comes from a dynamic traffic assignment (see section 4.3.3), and thus reflects realistic congestion patterns. Public transit uses a so-called 'matrix-based public transit', which uses matrices of attributes such as travel time or distance between public transit stops (see http://matsim.org/matrixbasedptrouter). The travel times for travelling by bicycle or on foot are computed by taking the travel distance with a constant velocity of 15 km/h (bicycle) or 5 km/h (walking).

In all cases, the generalized cost of travel is computed according to the same principles as in the dynamic traffic assignment; in the current implementation the utility weights are also the same. One consequence of this is that the accessibility in fact refers to a certain time of day. Technically the least cost path tree computation, anchored at the origin i, starts at a specific time of day and then executes a time-dependent Dijkstra shortest path algorithm (Lefebvre and Balmer, 2007). Accessibility of the same location at a different time of day will usually be different, since congestion patterns are different.

4.2.3 Spatial Resolution

When looking at high-resolution accessibility calculations, there are, in fact, two resolutions to consider: one that defines for how many origins i the accessibility is to be computed; and a second one that defines to what level the opportunities k are to be resolved.

Spatial resolution of the origins

In the present implementation, the origin side can be calculated for two spatial units: cells or zones:

- Cell-based approach: in this approach the study area is subdivided into square cells, where the resulting cell centroids serve as origins or measuring points for the accessibility calculation; see Figure 4.2(a). The spatial resolution depends on the selected cell size, which is configurable.
- Zone-based approach: this approach uses zone centroids as measuring points; see Figure 4.2(b). The centroid coordinates can be obtained from a variety of definitions. In this chapter, they are determined by averaging all parcel coordinates that belong to a zone. This corresponds to weighting each parcel equally; this may not be justified when, say, the number of residents or households varies strongly between parcels. The number of measuring points is defined by the number of zones.

In both cases, the accessibility computation is valid for the measuring point. In the grid-based approach it is then assumed that accessibility for

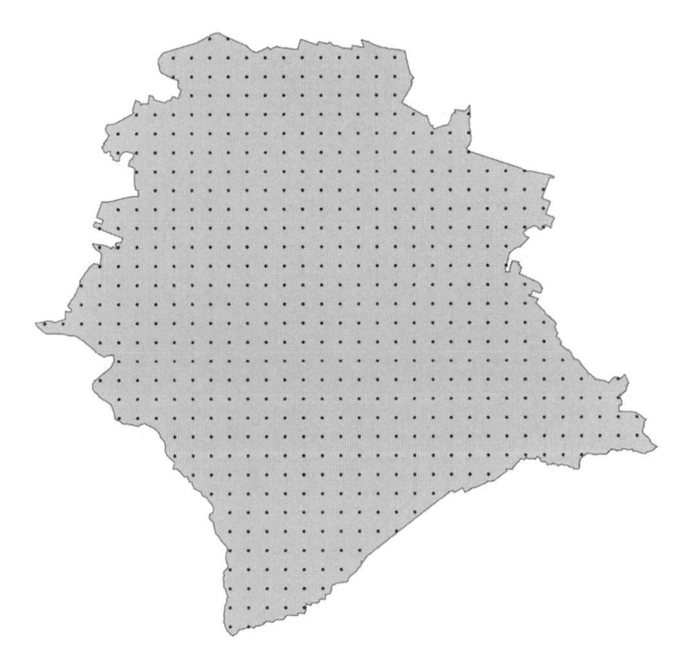

Figure 4.2 Cell- and zone-based approach in accessibility calculation for the example of the city of Zurich (grey area)
(a) Cell-based approach for the example of 400 m×400 m side length

in-between locations should be interpolated. In contrast, for the zone-based approach the accessibility of the measuring point is taken as representative for the whole zone. In consequence, the choice of the procedure of how to generate the measuring point for the zone-based approach has an influence on the results. There is no corresponding choice for the grid-based approach, removing one element of arbitrariness compared to the zone-based approach.

The following paragraphs concentrate on the cell-based approach. Nevertheless, the calculation procedure of the logsum term is the same for both approaches.

Spatial resolution of opportunities

Opportunity locations such as workplaces are given by land use. As stated earlier, they are attached to the nearest network node. The fact that they are attached to the nearest network node rather than the nearest network element (that is, including links) is, in fact, the only approximation that the present chapter makes with respect to the spatial resolution of opportunities.

Figure 4.2 (b) Zone-based approach

In Figure 4.3(a) opportunity locations (dots) provided by the land-use model are at a disaggregated parcel level. The spatial resolution inside MATSim depends on the resolution of the road network, that is, on the number of nodes and link lengths. Thus opportunities are directly aggregated to their nearest node on the given road network as depicted in Figure 4.3(b).

4.2.4 Computational Procedures

Exploring the entire network by using the 'least cost path tree' is a computationally expensive task. In order to accelerate the overall computing speed, the execution time of the least cost path tree is reduced by the following elements.

Origins

For each origin location i, the nearest node on the road network is identified. Locations that share the same node have the same travel disutilities on the network. In this case the least cost path tree is executed only once and the calculated disutilities on the network are reused for all i that are mapped on the same node. Only the calculation of the travel disutility to

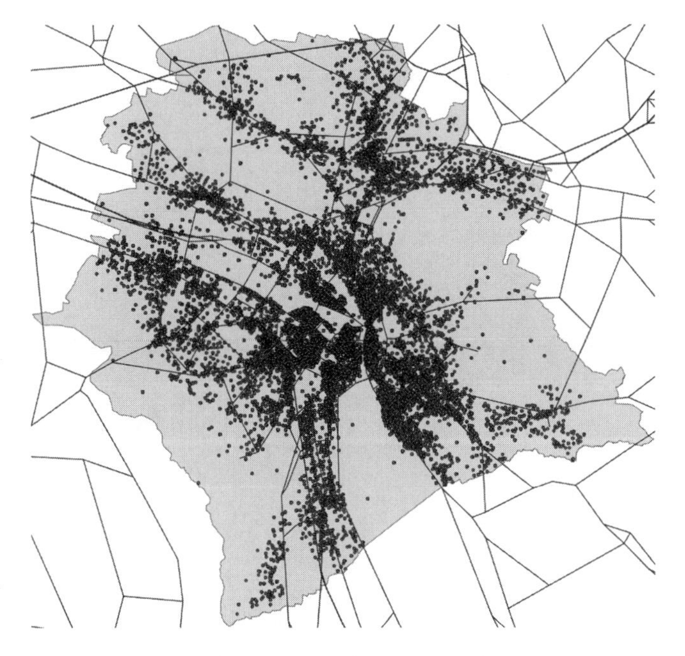

Figure 4.3 Opportunity locations and nearest node
(a) Opportunities with parcel coordinates given by the land-use model

overcome the gap between location i and the network is done individually. This implies the approximation that all origins assigned to the same node use the same time of day as starting time for the time-dependent least cost path tree algorithm.

Opportunities

In fact it is sufficient to sum over all opportunities k attached to a node j only once. For this, assume that the travel disutility V_{ik} can be decomposed as:

$$V_{ik} = V_{ij} + V_{jk} \quad \forall k \in j, \tag{4.8}$$

where the notation $k \in j$ shall refer to all opportunities k attached to node j. Then:

$$\sum_{k \in j} e^{V_{ik}} = \sum_{k \in j} e^{(V_{ij}+V_{jk})} = \sum_{k \in j} e^{V_{ij}} e^{V_{jk}} = e^{V_{ij}} \sum_{k \in j} e^{V_{jk}} =: e^{V_{ij}} \cdot Opp_j. \tag{4.9}$$

Thus, it is sufficient to compute Opp_j once for every network node j and from then on to compute accessibilities by:

Figure 4.3 (b) *Opportunities, aggregated on the nearest node on the road network*

$$A_i = \ln \sum_k e^{V_{ik}} = \ln \sum_j e^{V_{ij}} \cdot Opp_j. \tag{4.10}$$

This is an exact result. The only approximation was done earlier; it is the assumption that all opportunities are reached through the nearest network node rather than directly from the links.

4.3 SCENARIO: ZURICH, SWITZERLAND

The above approach is applied to a real-world scenario. This is the city of Zurich, a parcel-based UrbanSim application that will be briefly discussed here. A full description is given by Schirmer et al. (2011) and Schirmer (2010).

The Zurich application consists of 40407 parcels, 336291 inhabitants and 316703 jobs. In this chapter the UrbanSim base year 2000 is used to create the input for the MATSim runs. After that UrbanSim is no longer needed for the present study.

4.3.1 Population and Travel Demand

In order to speed up computation times, MATSim considers a 10 per cent random sample of the synthetic UrbanSim population, consisting of 33 629 agents. All MATSim agents have complete day plans with 'home-to-work-to-home' activity chains. Work activities can be started between 7 and 9 a.m. and have a typical duration of eight hours. The home activity has a typical duration of 12 hours and no temporal restriction.

4.3.2 Network and Adjustments

A revised Swiss regional planning network is used (Vrtic et al., 2003; Chen et al., 2008) that includes major European transit corridors; see Figure 4.4. The network consists of 24 180 nodes and 60 492 links, where each link is defined by an origin and a destination node, a length, a free speed car travel time, a flow capacity and a number of lanes. In addition, each link obtains congested car travel times once the traffic flow simulation in MATSim is completed (see Nicolai and Nagel, 2012).

The flow and storage capacities of the road network are automatically adjusted based on the given population sampling rate used for the MATSim runs. This is done in order to preserve congestion effects when running MATSim at small samples. The flow capacity gives the maximum number of vehicles per time unit that can pass a link (Cetin et al., 2003). It is adjusted by a flow capacity factor, which is set to the same value as the given *Population Sampling Rate*. The storage capacity defines the maximum number of vehicles that can be on a link (Cetin et al., 2003). The corresponding storage capacity factor is defined as *Population Sampling Rate/Heuristic Factor*, where the *Heuristic Factor* = *Population Sampling*

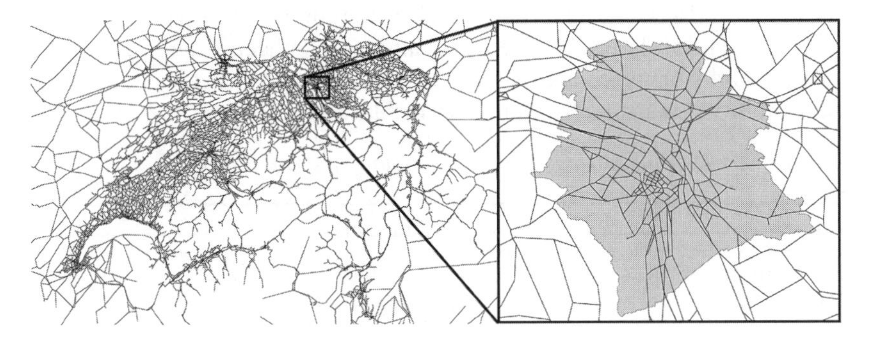

Figure 4.4 The Zurich case study network, area of Zurich (light grey) enlarged

*Rate*1/4. The *Heuristic Factor* aims to raise the storage capacity especially at low sampling rates to avoid network breakdowns caused by strong but spurious backlogs (Rieser and Nagel, 2008).

4.3.3 Traffic Simulation

First, a base case MATSim run is performed by running the simulation for 1000 iterations. During the first 800 iterations 10 per cent of the agents perform 'time adaptation', which changes the departure times of an agent, and 10 per cent adapt their routes. The remaining agents switch between their plans. During the last 200 iterations time and route adaptations are switched off; thus, agents only switch between existing plans.

4.4 APPLICATION

This section will specifically look at the proposed high resolution accessibility approach using the example of workplace accessibility. In particular the influence of the spatial resolution on the quality of the results and on computational performance are considered. In addition, as one example of a sensitivity test, the congested accessibility field is compared to uncongested accessibility as well as accessibility by bicycle and by walking.

All measurements are applied for the morning peak hour at 8 a.m. At this time most travellers are commuting to work. Table 4.1 summarizes relevant parameter settings.

4.4.1 Default Setting

Figure 4.5 depicts the accessibility outcome using the 'default setting' as stated in Table 4.1. To improve interpretability, the road network is overlaid. The scale bar on the right-hand side indicates the accessibility level. Good workplace accessibility is indicated by white areas and poor accessibility is indicated by dark grey or black areas.

Table 4.1 Default settings for the accessibility computation

Parameter	Default setting
Resolution	100 m × 100 m
Travel cost	Congested car
	Travel times (minutes)
β_{ttcar}	−12 utils/hour
β_{ttwlk}	−12 utils/hour

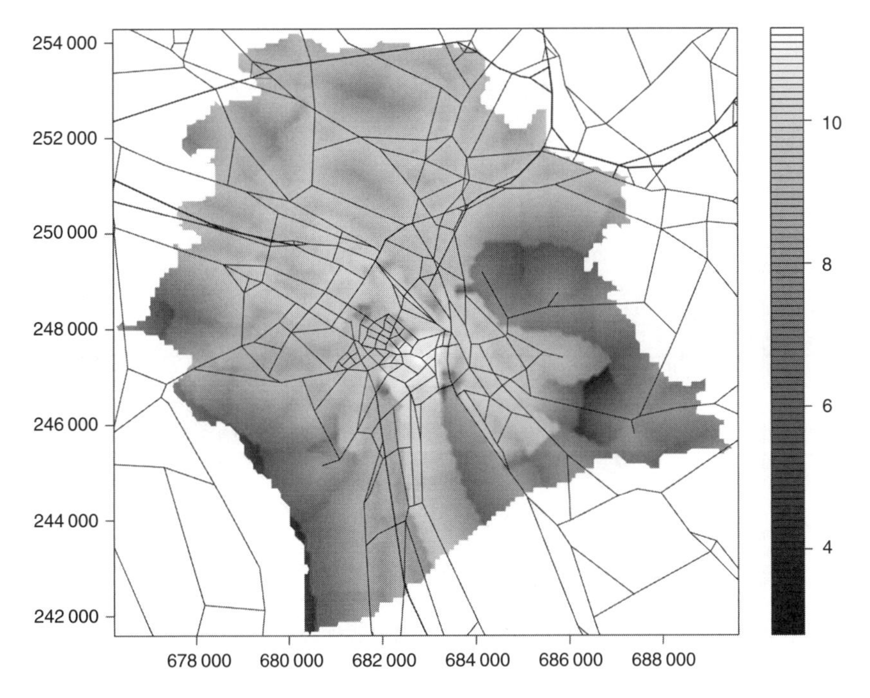

Figure 4.5 Outcome of the proposed accessibility measurement using the 'default settings' in Table 4.1

The plot exhibits very good workplace accessibility in areas that provide a high density of opportunities and a well-developed road network. These characteristics apply for the inner city, where the highest accessibility values are measured, and for the areas along the major access roads from and to Zurich, visible as white or light grey corridors. In contrast, areas with less workplace accessibility have a gradient from dark grey to black. This for instance applies to the 'Zürichberg' and the 'Uetliberg', which are two undeveloped wooded hills located in the eastern and south-western part of Zurich. Several 'islands of low accessibility' in the centre of Zurich are due to localized congestion on those links: if there is strong congestion at the origin, then all opportunities k incur a strongly negative V_{ik} and thus make only a small contribution to the sum.

It should be noted that accessibility is now smooth in space; see Figure 4.6. Figure 4.6(a) is the same as Figure 4.5 but without the road network. It is clear how highly accessible areas trace the road network. The zone representation of the same data is shown in Figure 4.6(b). Clearly all additional spatial resolution within the zones is now gone.

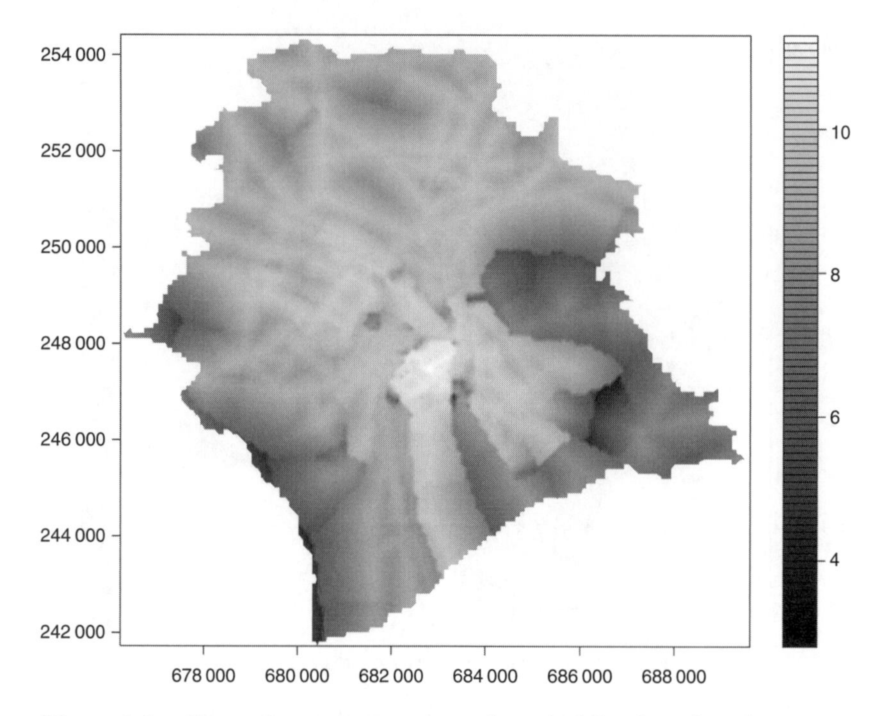

Figure 4.6 Alternative representations of accessibility based on the same congestion data as Figure 4.5
(a) The same as Figure 4.5, but without showing the road network

Also, with the zones approach the question of how far the transport network becomes visible in the accessibility plot hinges entirely on the question of whether the zones reflect the structure of the transport network or not. Such arbitrariness is removed with an approach that does not rely on zones.

4.4.2 Spatial Resolution

Figure 4.7 shows the outcome of the accessibility measurement at the following resolutions: 50 m × 50 m, 100 m × 100 m, 200 m × 200 m and 400 m × 400 m. In terms of information gain it can be stated that the lowest resolution with 400 m × 400 m provides a rather undifferentiated picture, whereas higher resolutions lead to more detailed measurements. In the 50 m × 50 m resolution, even fine road structures in the city centre are clearly visible. However, a significant increase in the level of detail can be observed up to the resolution of 100 m × 100 m. The higher resolution

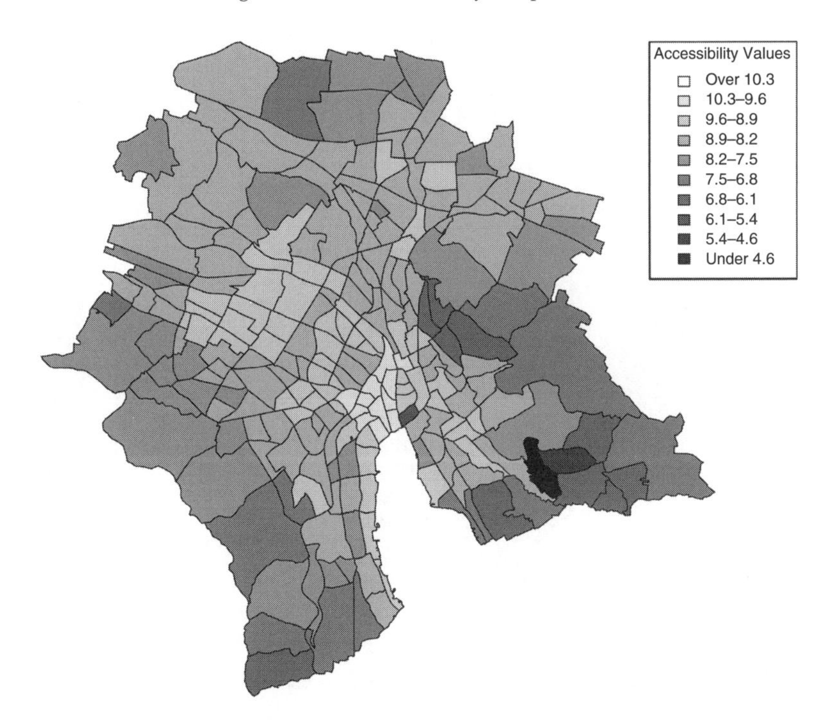

Figure 4.6 (b) *Accessibility based on the same congestion data as Figure 4.5, but based on zones as is traditionally done*

(50 m × 50 m) looks smoother and sharper, but does not offer noticeable gains.

4.4.3 Computing Times

In section 4.4.2, the resolution of two successive plots is doubled. This corresponds to a quadrupling of the measuring points. However, this increase is not reflected in the computing times (see Table 4.2), due to the run time optimizations outlined in section 4.2.4. Instead there is no advantage in terms of computing speed for the traditional zone-based accessibility measure despite the low resolution.

All measurements are performed on a Mac Book Pro with an Intel Core 2 Duo 2.5GHz processor and 4 GB of memory. Currently one CPU core is used to execute the accessibility computations. Computing times could be further improved by using multiple threads: Since the computations for different origins are independent, they could be distributed between all available CPU cores.

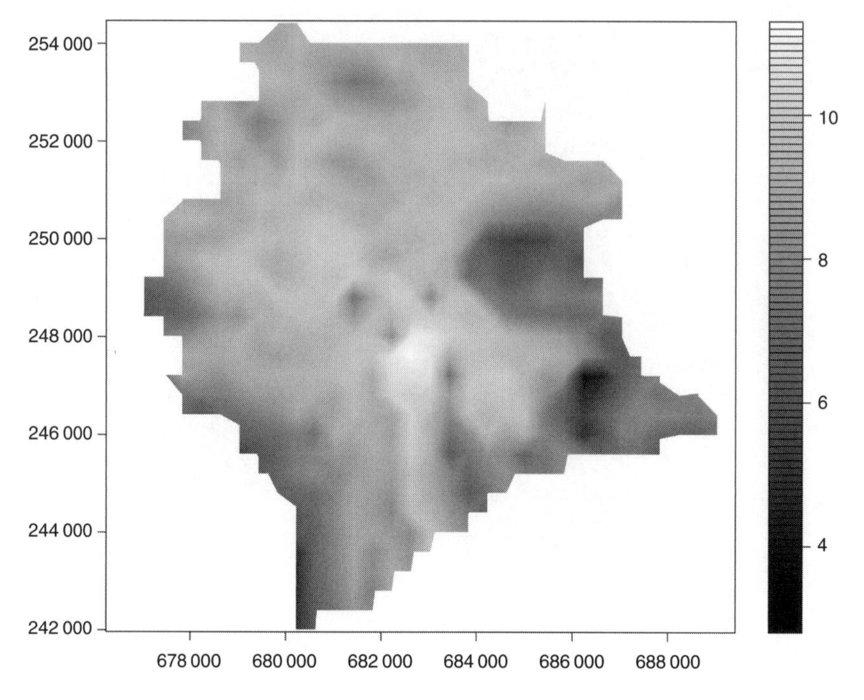

Figure 4.7 Results of different resolutions. In ascending order with the following cell sizes 400 m × 400 m, 200 m × 200 m, 100 m × 100 m and 50 m × 50 m

(a) Resolution 400 m × 400 m

Table 4.2 Computation times to measure accessibility at different resolutions and the zone level

Cell resolution	Origins	Aggregated opportunities	Computing time (min)
50 m × 50 m	36748	272	2–3
100 m × 100 m	9195	272	2
200 m × 200 m	2292	272	≈2
400 m × 400 m	577	272	≈1
Zone resolution	**Origins**	**Aggregated opportunities**	**Computing time (min)**
Given by zones	234	272	≈1

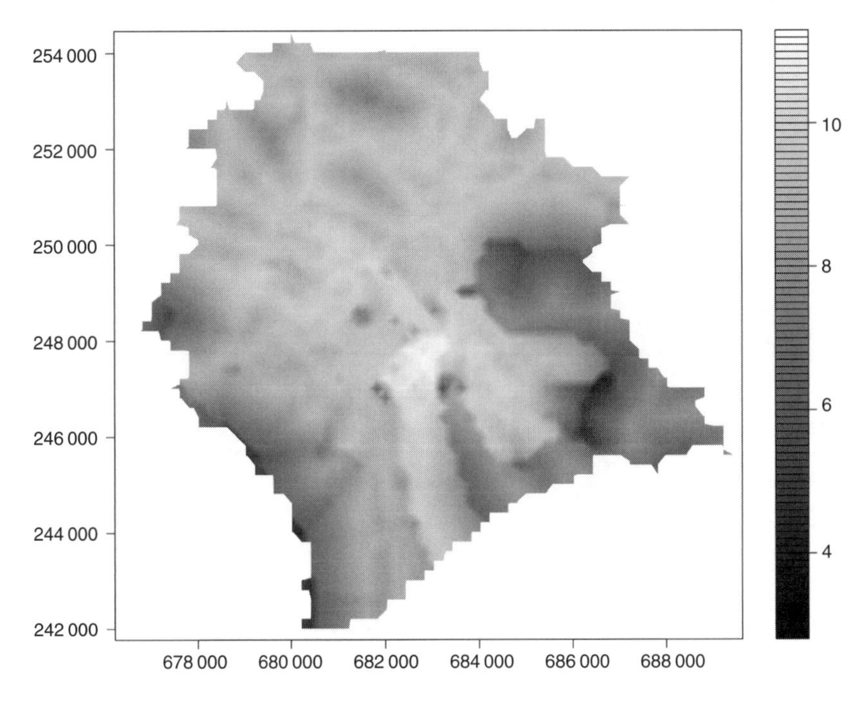

Figure 4.7 (b) Resolution 200 m × 200 m

4.4.4 Mode

The results of congested versus uncongested car travel and of using bicycle or walk mode are shown in Figure 4.8. The first two plots illustrate how congestion significantly reduces accessibility.

The third and fourth plots illustrate accessibility by bicycle and by walking, respectively. Congestion has no effect on these modes. Both plots show that spatial proximity to opportunities has a strong influence on the results. Locations away from the city centre and the commercial areas in the northeast and western part of Zurich have rapidly decreasing accessibility. As seems plausible, near the city centre the bicycle provides similar accessibility to the car under congested conditions. Farther away from the centre, the car gains ground, even under congested conditions.

One might speculate that Figure 4.8(a) is what people who plan to use a car expect, but Figure 4.8(b) is what they actually get. And what they actually get is not much different from what one gets by using a bicycle (Figure 4.8(c)). In contrast, walking is much worse (Figure 4.8(d)).

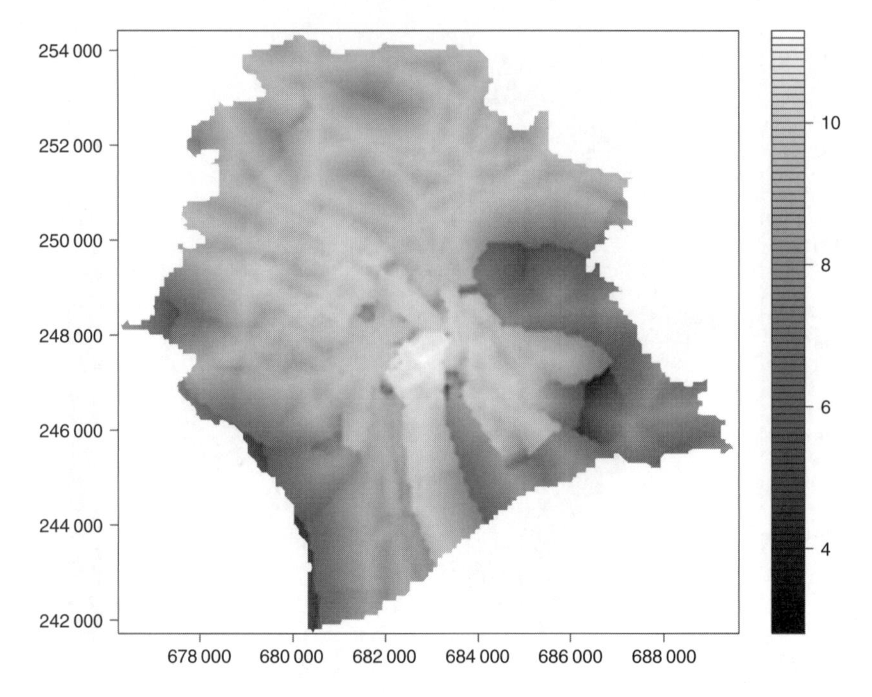

Figure 4.7 (c) *Resolution 100 m × 100 m*

4.5 DISCUSSION

4.5.1 The Unit of the Accessibility Measurement

The logsum term, equation (4.4), sometimes contains a scale parameter, that is, it reads $(1/\mu)\, ln \sum_k exp(\mu \widetilde{V}_{ik})$(*). When the utility function is estimated, equation (4.4) is the correct form (see also Train, 2003). Clearly the user can decompose the estimated utility function V_{ik} as $ik = \mu\widetilde{V}_{ik}$, in which case equation (4.4) needs to be replaced by equation (*) in order to obtain the accessibility in the rescaled units. This may be useful for example if the utility function is to be scaled to monetary units and accessibility is to be expressed in those same monetary units. Since $\mu \widetilde{V}_{ik}$ is the same as V_{ik}, such rescaling does not change the structure of the plots; however, all accessibility values will be multiplied by $1/\mu$.

4.5.2 Spatial Resolution

Most plots in this chapter are based on a grid resolution of 100 m × 100 m and the road network as shown in some of the figures. As it was argued,

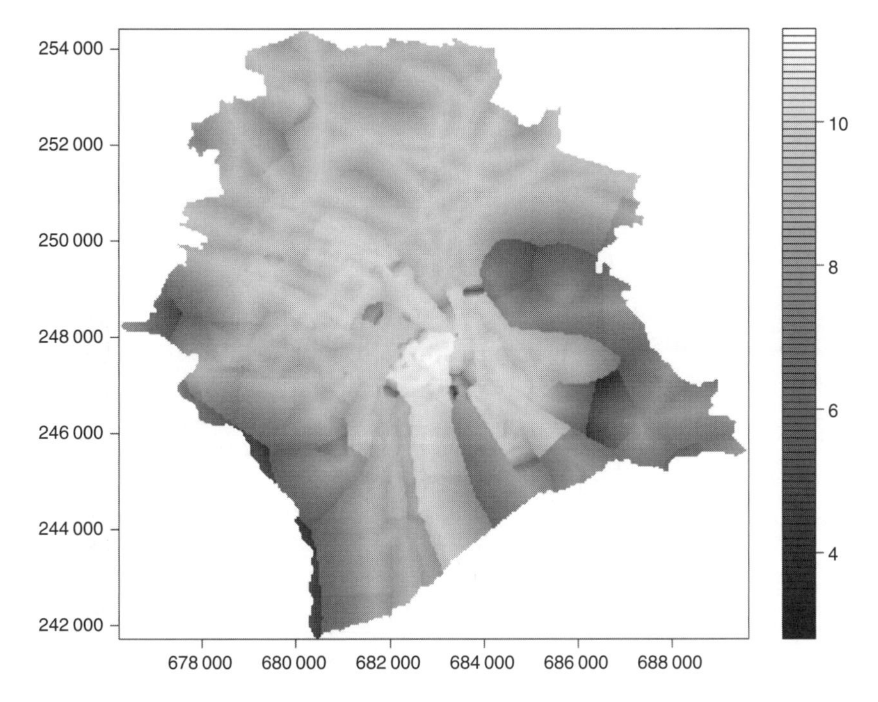

Figure 4.7 (d) *Resolution 50 m × 50 m*

higher spatial resolutions with the same road network do not lead to discernible improvements. In contrast, a higher resolution road network would clearly make a difference (see also below). Using a higher resolution road network is feasible; the computational cost scales roughly in the number of links since the worst case complexity of the Dijkstra tree computation is roughly linear in the number of links for planar graphs. In the present situation, it was decided to stick with the lower resolution network. One reason was that we wanted to investigate how our approach would perform in such a situation. Another reason was that there was no higher resolution network together with a calibrated scenario available.

4.5.3 Connection to the Transport Network

In the present investigation, origins are connected by the walk mode to the nearest network element (node or link) and destinations are connected by the walk mode to the nearest network node. Neither of these assumptions is always correct. For example in the Zurich case there are additional roads beyond those that are used for the computation and thus it is quite

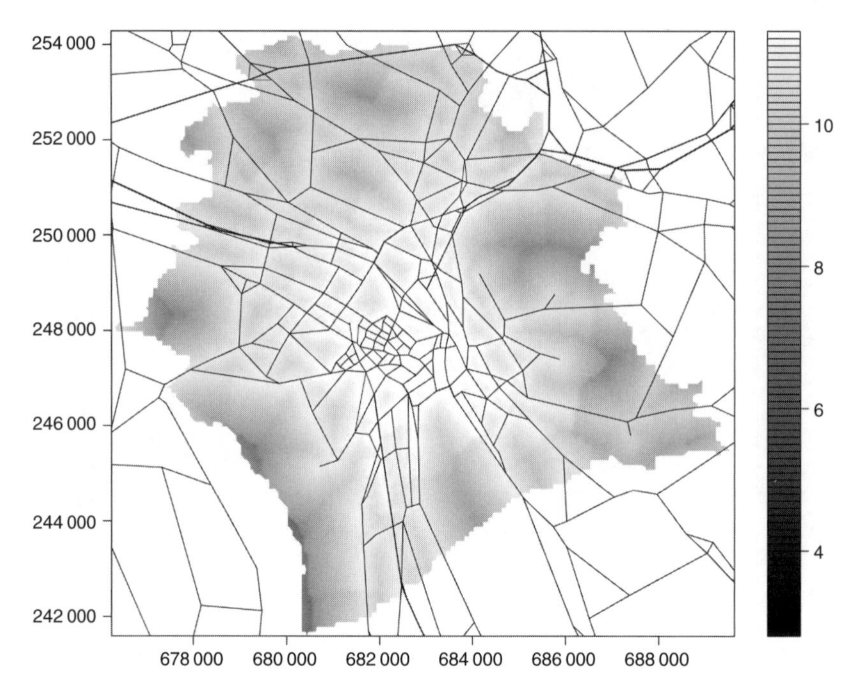

Figure 4.8 Plots visualizing the influence of different transport modes on the accessibility computation

(a) Free speed car travel times in accessibility computation

plausible that both for the car and for the bicycle modes access to the network is much faster than assumed in the plots. This would reduce the 'accessibility deprivation' that is currently visible (Figure 4.5) in areas far away from the network. However, not all areas are in fact covered by such a secondary network, which makes it difficult to come up with a general specification for such areas, and which is the reason why the 'walk' assumption was made.

Another issue is that origins are attached to the nearest network element, rather than to the network element that offers the highest level of accessibility. This sometimes causes seemingly sharp accessibility changes far away from the transport network, for example around the coordinates (685000, 249500) in Figure 4.5. It might seem plausible to give origins a choice to which network element they want to attach, rather than forcing their attachment by such an arbitrary algorithm. On the other hand, in reality parcels are attached to the transport network by specific links and in consequence one should not invent better algorithms but use better data. Also, the area around the above-mentioned coordinates is, in fact, a

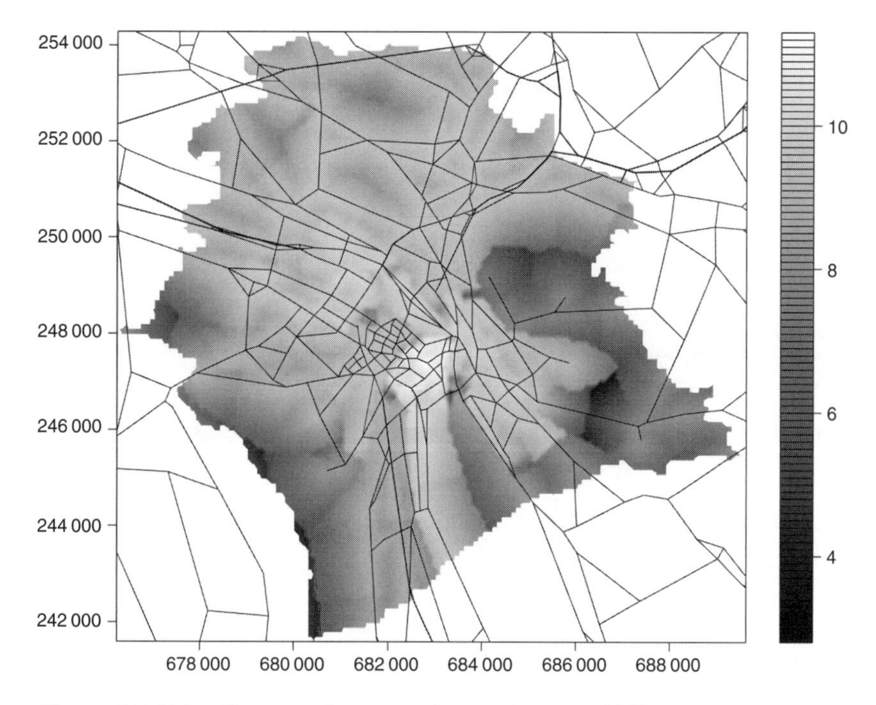

Figure 4.8 (b) *Congested car travel times in accessibility computation (reference setting)*

communal forest and thus the issue is, albeit not visually pleasing, of little practical relevance. It is important to note that both issues are significantly reduced when moving to a higher resolution network, since any kind of approximation about how to reach the transport network will then have a smaller impact.

4.5.4 Directionality

The introduction differentiated between 'accessibility of a location to other locations that provide services' and 'accessibility of a location providing services from other locations'. The actual computations have concentrated on the first variant: a time-dependent shortest path tree was expanded from the origin location. The resulting measure could also be interpreted as the accessibility of the location from other locations if travel effort was symmetric, that is, if the effort to go from A to B was the same as the effort of going from B to A. Since the computations refer to a specific time of day, this will in general not be the case: for example, during

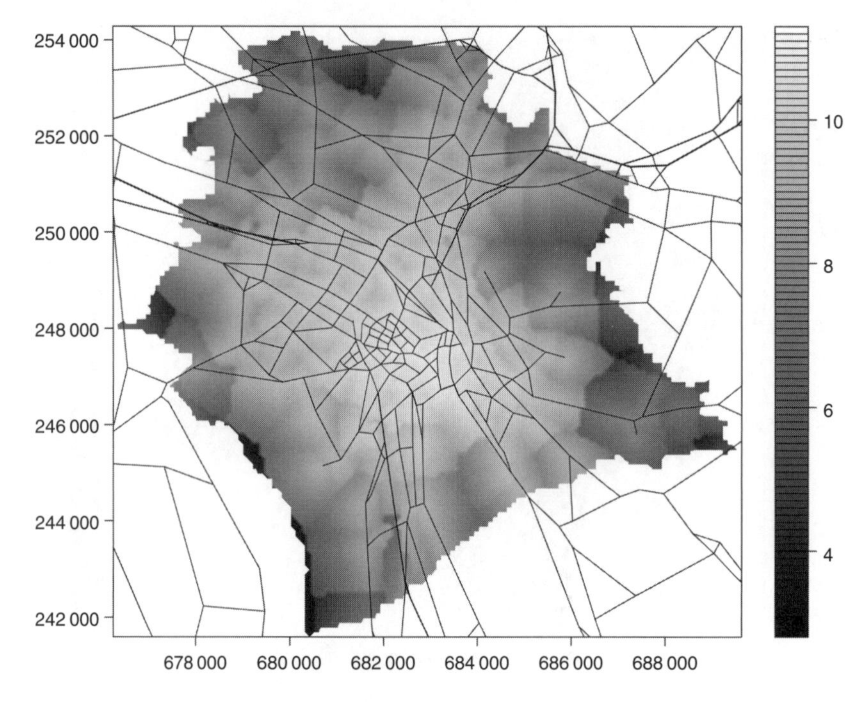

Figure 4.8 (c) Bicycle travel times in accessibility computation

the morning peak there is probably more congestion into the centre than going out. That is, 'origin accessibility' and 'destination accessibility' at best approximate each other.

However, the method described in the present chapter could be modified to compute a high resolution destination accessibility, starting at the destination and expanding a backwards shortest path tree, anchored at a specific arrival time.

4.5.5 Possible Applications

The concept of accessibility has always been used to analyse the 'value' of locations. Moving it away from the zone and onto a point clarifies that it is really a 'point' concept. As an important result, zones are no longer needed; accessibility can be computed for regions for which zones are not available; and all arbitrariness related to the definition of zones is removed. The approach also works equally well for slow modes such as walk and bicycle; only the network data needs to be more detailed.

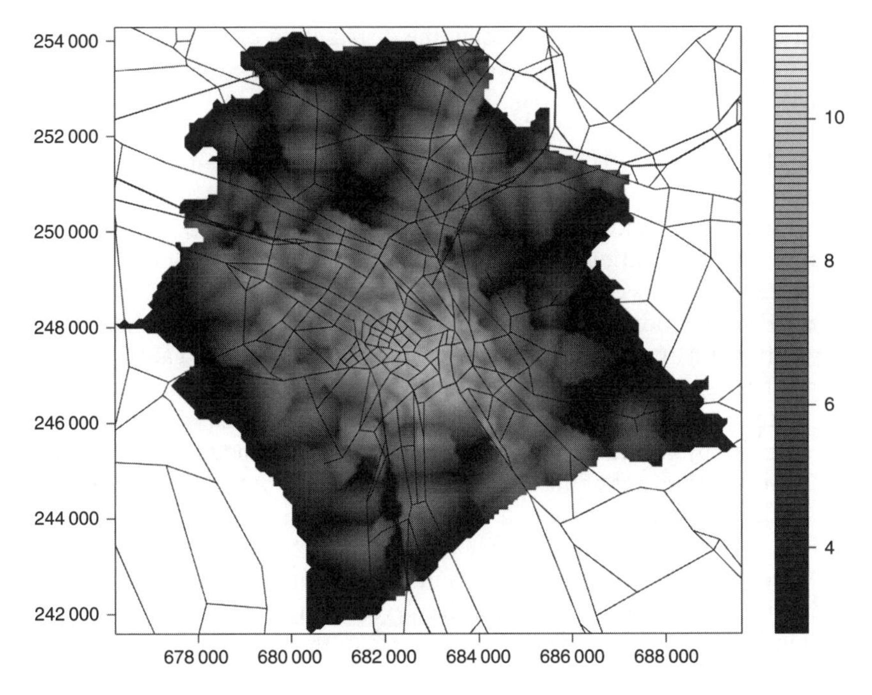

Figure 4.8 (d) *Walk travel times in accessibility computation*

Besides helping with location choice, the approach can also be used to systematically and automatically assess the access of locations to services, for example to schools, hospitals, groceries and the like. For this the locations of these opportunities need to be known, but freely available data sources such as OpenStreetMap (http://openstreetmap.org) increasingly make such data available, or could be retrofitted even by a community effort if desired. Clearly also the impedance function $f(c_{ik})$ needs to be adapted in a meaningful way to the measure of interest.

4.6 CONCLUSION

Accessibility measures are, for a given origin, a weighted sum over possible destinations, where the weight typically is a decreasing function of generalized cost, possibly multiplied with an attractiveness indicator. In this chapter, the econometric logsum term is used as an example. It can also be interpreted as a benefit that someone at a specific location derives from access to spatially distributed opportunities. It includes, as usual, a land-use component by considering the distribution of opportunities and a transport component by determining the effort to get there. Important results include:

- The approach is able to provide a spatially differentiated picture. Artefacts such as supposedly homogeneous accessibility within zones are removed.
- The computation of a cell-based accessibility measurement at high resolutions is computationally feasible.
- For the present scenario and network a resolution finer than 100 m × 100 m does not deliver noticeable additional gains.
- The effect of congestion is clearly visible. The result also confirms that in the urban core accessibility by bicycle is similar to accessibility by car during congested peak hours. Outside the urban core, however, accessibility by bicycle is clearly worse than by car also during peak hours.

In contrast, accessibility when walking is definitely much worse than either by car or by bicycle and reaches levels comparable to accessibilities by bicycle or by car only in the inner centre of the city.

ACKNOWLEDGMENTS

This work has been funded in part by the EU Seventh Framework Programme (FP7) within the research project SustainCity. The code for the calculation and visualization of accessibilities is based on previous work by Johannes Illenberger. Michael Balmer has programmed the efficient and flexible 'least cost path tree' algorithm that is at the heart of the accessibility computation. Michel Bierlaire provided access to the Zurich data. Michael Zilske was significantly involved in the runtime optimization of the accessibility calculation. Theresa Thunig implemented the interpolation routine.

NOTE

1. For example, the number of shopping opportunities within travel time t^* can be obtained by summing over all shopping opportunities j, $a_j = 1$ (all shopping locations equally attractive), $g(X) = X$, and:

$$f(c_{ij}) = \begin{cases} 1 & if\ t_{ij} \leq t^* \\ 0 & else \end{cases} . \quad (4.2)$$

REFERENCES

Balmer, M., B. Raney and K. Nagel (2005), 'Adjustment of activity timing and duration in an agent-based traffic flow simulation', in H.J.P. Timmermans (ed.), *Progress in Activity-Based Analysis*, Oxford: Elsevier, pp. 91–114.

Balmer, M., M. Rieser, K. Meister, D. Charypar, N. Lefebvre, K. Nagel and K.W. Axhausen (2009), 'MATSim-T: architecture and simulation times', in A.L.C. Bazzan and F. Klügl (eds), *Multi-Agent Systems for Traffic and Transportation*, Hershey, PA: IGI Global, pp. 57–78.

Ben-Akiva, M. and S.R. Lerman (1985), *Discrete Choice Analysis*, Cambridge, MA: MIT Press.

Bono, F. and E. Gutiérrez (2011), 'A network-based analysis of the impact of structural damage on urban accessibility following a disaster: the case of the seismically damaged Port Au Prince and Carrefour urban road networks', *Journal of Transport Geography*, **19** (6), 1443–1455; http://www.sciencedirect.com/science/article/pii/S0966692311001372.

Borzacchiello, M.T., P. Nijkamp and E. Koomen (2010), 'Accessibility and urban development: a grid-based comparative statistical analysis of Dutch cities', *Environment and Planning B: Planning and Design*, **37**, pp. 148–169; http://econpapers.repec.org/RePEc:dgr:vuarem:2007-15.

Cetin, N., A. Burri and K. Nagel (2003), 'A large-scale agent-based traffic microsimulation based on queue model', in *Proceedings of the Swiss Transport Research Conference (STRC)*, Monte Verita, CH; www.strc.ch. Earlier version, with inferior performance values: Transportation Research Board Annual Meeting, 2003, paper number 03–4272.

Chen, Y., M. Rieser, D. Grether and K. Nagel (2008), 'Improving a large-scale agent-based simulation scenario', VSP Working Paper 08–15, TU Berlin, Transport Systems Planning and Transport Telematics, www.vsp.tu-berlin.de/publications.

Condeço-Melhorado, A., J. Gutiérrez and J.C. García-Palomares (2011), 'Spatial impacts of road pricing: accessibility, regional spillovers and territorial cohesion', *Transportation Research Part A: Policy and Practice*, **45** (3), 185–203; http://www.sciencedirect.com/science/article/pii/S0965856410001618.

de Jong, G., A. Daly, M. Pieters and T. van der Hoorn (2007), 'The logsum as an evaluation measure: review of the literature and new results', *Transportation Research Part A: Policy and Practice*, **41** (9), 874–889; http://www.sciencedirect.com/science/article/pii/S0965856407000316.

Dijkstra, E. (1959), 'A note on two problems in connexion with graphs', *Numerische Mathematik*, **1**, 269–271.

Fröhlich, P. and K.W. Axhausen (2004), 'Sensitivity of accessibility measurements to the underlying transport network model', IVT Working paper 245, Institute for Transport Planning and Systems, ETH Zurich, Zurich; http://www.ivt.ethz.ch/vpl/publications/reports.

Frost, M.E. and N.A. Spence (1995), 'The rediscovery of accessibility and economic potential: the critical issue of self-potential', *Environment and Planning A*, **27** (11), 1833–1848.

Geurs, K.T. and J.R. Ritsema van Eck (2001), 'Accessibility measures: review and applications', Technical report, National Institute of Public Health and the Environment, Bilthoven, June.

Geurs, K.T. and B. van Wee (2004), 'Accessibility evaluation of land-use and transport strategies: review and research directions', *Journal of Transport Geography*, **12**, 127–140.

Gutiérrez, J., A. Condeço-Melhorado and J.C. Martín (2010), 'Using accessibility indicators and GIS to assess spatial spillovers of transport infrastructure investment', *Journal of Transport Geography*, **18** (1), 141–152; http://www.sciencedirect.com/science/article/pii/S096669230800152X.

Gutiérrez, J. and G. Gómez (1999), 'The impact of orbital motorways on intra-metropolitan accessibility: the case of Madrid's M-40', *Journal of Transport Geography*, **7** (1), http://www.sciencedirect.com/science/article/pii/S0966692398000295.

Hansen, W. (1959), 'How accessibility shapes land use', *Journal of the American Planning Association*, **25** (2), 73–76.

Kwan, M-P. (1998), 'Space–time and integral measures of individual accessibility: a comparative analysis using a point-based framework', *Geographical Analysis*, **30** (3), 191–216; http://dx.doi.org/10.1111/j.1538-4632.1998.tb00396.x.

Lefebvre, N. and M. Balmer (2007), 'Fast shortest path computation in time-dependent traffic networks', in *Proceedings of the Swiss Transport Research Conference (STRC)*, Monte Verita, CH, September; see www.strc.ch.

Liu, S. and X. Zhu (2004), 'Accessibility analyst: an integrated GIS tool for accessibility analysis in urban transportation planning', *Environment and Planning B: Planning and Design*, **31**, 105–124.

Martín, J.C. and A. Reggiani (2007), 'Recent methodological developments to measure spatial interaction: synthetic accessibility indices applied to high-speed train investments', *Transport Reviews*, **27** (5), 551–571; http://www.tandfonline.com/doi/abs/10.1080/01441640701322610.

Miller, E., K. Nagel, H. Ševčíková, D. Socha and P. Waddell (2005), 'OPUS: an open platform for urban simulation', in *9th Conference on Computers in Urban Planning and Urban Management (CUPUM)*, University College London; http://128.40.111.250/cupum/searchpapers/detail.asp? pID=428; see www.cupum.org.

Moeckel, R. (2006), 'Business location decisions and urban sprawl: a microsimulation of business relocation and firmography', PhD thesis, University of Dortmund.

Nicolai, T.W. and K. Nagel (2012), 'Coupling transport and land-use: investigating accessibility indicators for feedback from a travel to a land use model', in *Latsis Symposium 2012– 1st European Symposium on Quantitative Methods in Transportation Systems*, Lausanne, Switzerland; also VSP WP 12–16, see www.vsp.tu-berlin.de/publications.

OPUS User Guide (2011), *The Open Platform for Urban Simulation and UrbanSim Version 4.3*, University of California Berkeley and University of Washington, January 2011; http://www.urbansim.org.

Raney, B. and K. Nagel (2006), 'An improved framework for large-scale multi-agent simulations of travel behaviour', in P. Rietveld, B. Jourquin and K. Westin (eds), *Towards Better Performing European Transportation Systems*, London: Routledge, pp. 305–347.

Rieser, M. and K. Nagel (2008), 'Network breakdown "at the edge of chaos" in multi-agent traffic simulations', *European Journal of Physics*, **63** (3), 321–327.

Schirmer, P. (2010), 'Options and constraints of a parcel based approach in UrbanSimE', in *Proceedings of the 10th Swiss Transport Research Conference (STRC)*; see www.strc.ch.

Schirmer, P., C. Zöllig, B.R. Bodenmann and K.W. Axhausen (2011), 'The Zurich case study of UrbanSim', *European Regional Science Association Conference*, June.

Train, K. (2003), *Discrete Choice Methods with Simulation*, Cambridge: Cambridge University Press.

Vandenbulcke, G., T. Steenberghen and I. Thomas (2009), 'Mapping accessibility in Belgium: a tool for land-use and transport planning?', *Journal of Transport Geography*, **17** (1), 39–53; http://www.sciencedirect.com/science/article/pii/S096669230800032X.

Vrtic, M., P. Fröhlich and K.W. Axhausen (2003), 'Schweizerische Netzmodelle für Strassen- und Schienenverkehr', in T. Bieger, C. Lässer and R. Maggi (eds), *Jahrbuch 2002/03 Schweizerische Verkehrswirtschaft*, St. Gallen: Schweizer Verkehrswissenschaftliche Gesellschaft, pp. 119–140.

Waddell, P. (2002), 'Urbansim: modeling urban development for land use, transportation, and environmental planning', *Journal of American Planning Association*, **68** (3), 297–314.

Zhu, X. and S. Liu (2004), 'Analysis of the impact of the MRT system on accessibility in Singapore using an integrated GIS tool', *Journal of Transport Geography*, **12** (2), 89–101; http://www.sciencedirect.com/science/article/pii/S096669230300058.

5. Sensing 'socio-spatio' interaction and accessibility from location-sharing services data

Laurie A. Schintler, Rajendra Kulkarni, Kingsley Haynes and Roger Stough

5.1 INTRODUCTION

Web 2.0 services and applications are contributing to an unprecedented and growing collection of digital data that are rich in detail about human behaviour. Location-sharing services, which allow individuals to 'check in' to locations via Global Positioning System (GPS)-equipped devices and then share this information with friends, are generating detailed accounts of the space–time movements of individuals and related patterns of social and spatial interaction. Each 'check-in' records the latitude and longitude coordinates of the location as well as the time of the transaction. Users can also provide a brief comment with their check-in. For the purpose of understanding human mobility patterns, location-sharing services data has some advantages over more traditional sources of data, such as that collected via travel demand surveys or activity-based diaries. First, the data captures not only the behaviour of residents within a region, but also that of visitors to the region, for example tourists or business travellers. Second, the format and content of the data is consistent across different regions, allowing for cross-city and regional comparisons. Third, the information contained in the data has a high degree of spatial and temporal resolution. Lastly, the data contains information on the social ties of individuals, which allows for the simultaneous consideration of digital socialization and spatial interaction.

One issue pertaining to location-sharing services data, and more generally volunteered geographic information (VGI), is sample bias (Elwood et al., 2012). Users of location-sharing services tend to be low-to-middle-income men between the ages of 19 and 29 years (Zickuhr, 2012). Users also differ in the frequency with which they report information. Thus, the complete set of movement patterns for certain individuals may be under-represented. On the other hand, there is some evidence to suggest that aggregate mobility patterns revealed from location-sharing services data are somewhat consistent with those derived from mobile phone transactions (Cho et al., 2011). Moreover use of location-sharing services, especially for smartphone owners, is on the rise. According to a recent study conducted by Pew, one in five smartphone users in the USA currently uses a location-sharing (or geo-social) service (Zickhur, 2012).

In the last few years, there have been significant efforts to mine location-sharing services data and other similar types of geo-social digital data, for example mobile phone trajectories, to characterize the complexity of different aspects of human mobility, although very few studies to date have examined how these patterns of mobility and spatial interaction vary within and across different regions. This chapter intends to fill this gap in the literature. Specifically, we use bipartite network modelling to derive a set of metrics for characterizing regional variations in the mobility patterns of individuals and the complexity of this phenomenon. Through this study we also attempt to gain an understanding of what types of trips location-sharing services data may represent. For example, are they short-distance or long-distance trips? Lastly, we apply a community detection algorithm to the bipartite network to derive 'mobility sheds', or regions in which the same groups of people tend to travel or access. For the purpose of this study we draw upon a sample of Brightkite location-sharing service user check-ins made over the period April 2008 to December 2010 (Cho et al., 2011).

5.2 LITERATURE REVIEW

Location-sharing services data and other types of geo-social data are leading to new insights into the complexity of human mobility. Recent research indicates that human movement exhibits certain spatial and temporal regularities, which when characterized in the aggregate are highly predictable. Based on mobile phone trajectories, scaling properties have been found in the population distributions of a number of aspects of mobility, including travel distances, time spent at locations, range of movement and even the popularity of places in terms of return visits. Surprisingly, these regularities do not appear to be an artefact of work schedules, but rather stem from intrinsic characteristics of human behaviour (Song et al., 2010; Gonzalez et al., 2008; Barabasi, 2007; Candia et al., 2008). Similar types of regularities and properties have been found based on location-sharing services (Cho et al., 2011; Cheng et al., 2011; Scellato and Mascolo, 2011; Noulas et al., 2011b).

Only a handful of studies have examined how these patterns and complexities vary within and across different regions. Noulas et al. (2011a) use Foursquare location-sharing services data to compare the average distance displacement of individuals across a set of world cities. They find an inverse relationship between average displacement and population density. In other words, dense cities like New York City have a much more limited range of movement than cities that are less dense. Cheng et al. (2011) conduct a similar study using Foursquare data; however, they use average 'radius of gyration' rather than the average distance displacement to capture the range of movement of individuals. If density, rather than distance, is the key component controlling interaction, this could provide support for the use of intervening opportunities as opposed to the distance-based hypothesis in traditional (gravity) interaction theory.

The statistics used in these studies have some limitations. First, they are vulnerable to outliers (and skewed distributions); given that mobility patterns exhibit scaling properties their use could be problematic. Second, they do not yield any information on how regions are directly and indirectly interconnected via the movement patterns of individuals. To what extent does a location share visitors with other locations? Further, how are locations connected, both directly and indirectly, via the individuals who check in to the location? Moreover, are locations with visitors in common spatially proximate or are they physically disjoint? A few studies have attempted to address the latter question. In particular, Lawlor et al. (2012) use a network-based community detection approach to identify regions in which the same individuals tend to circulate. They find that such communities do exhibit spatial association: that is, communities are concentrated in space. However, the size and structure of the communities were found to be spatially distinct.

This chapter builds upon the aforementioned research in a few respects. First, we develop a set of network-based metrics and related methods that can capture not only the mobility patterns of individuals (including their range of movement), but also the extent to which locations are connected via these displacements. Unlike range of gyration and average distance displacement, the indicators we develop are not vulnerable to statistical outliers. Second, we examine regional variations in displacement at a geographic scale that has not yet been done at the county level in the USA.

5.3 'SOCIO-SPATIO' BIPARTITE NETWORK

The indicators we formulate in this chapter are based on bipartite network modelling. A bipartite network (or two-mode network) contains vertices that can be partitioned into two disjoint sets. It includes only connections from vertices in one set to vertices in the other set. In other words, there are no connections within sets (Freeman and Duqueene, 1993). One of the earliest applications of bipartite graph theory is the Davis study, which looked at the participation of 15 women in 14 events in the southern USA (Davis et al., 1941). Since then, bipartite network modelling has been applied extensively in a wide array of contexts to examine such things as memberships in organizations, the structural properties of citation and co-authorship networks, and even to route structures in aviation networks. Very recently it has been used to examine structural patterns in space (e.g. Schmutte, 2014) and time (e.g. Greene and Cunningham, 2010).

We specify a bipartite graph with one set of vertices representing locations and the other individuals. For this study, we designate counties in the USA as the locations. An edge is present between a county and an individual if the person has visited that county. We do not consider the frequency with which an individual visits a single location, but rather only if they have been present at that place at any time. Thus, we formulate an unweighted two-mode network.

From any bipartite network, it is possible to extract two structurally equivalent one-mode networks (Wasserman and Faust, 1994). From the 'socio-spatio' bipartite network, we use the one-mode network which has counties as vertices and common users as edge weights to derive a set of county-level mobility indicators. Figure 5.1 provides a simple illustration

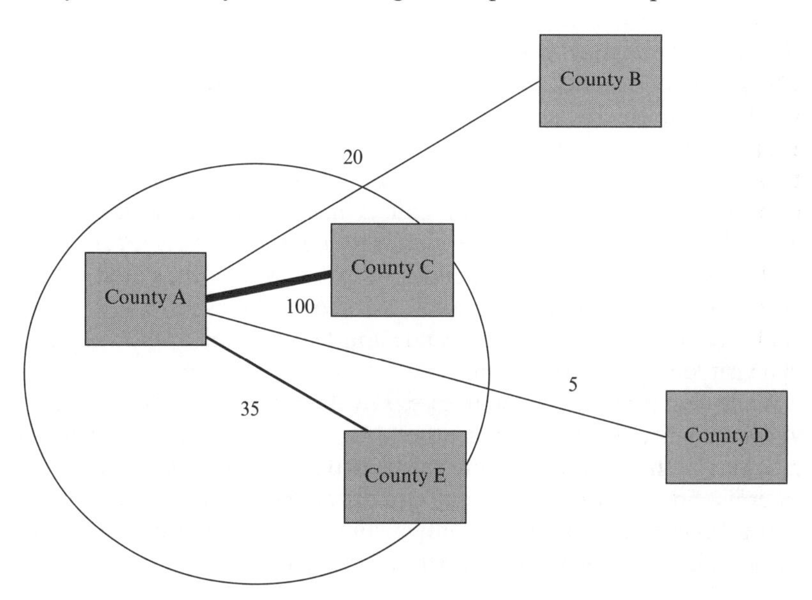

Figure 5.1 Simple illustration of how locations are connected in the one-mode network

of how counties are connected via this network structure. In this example County A is connected to Counties B, C, D and E through common visitors. For example, County A shares 20 unique visitors with County B. In other words, there were 20 individuals who visited both County A and County B. The counties contained in the circle represent counties that are in closer proximity to County A than those outside the circle.

In the one-mode network that we specify, vertex degree and weighted vertex degree provide two simple mobility indicators. In general, the degree of a vertex (or node) represents the number of other vertices to which that vertex is directly connected. Weighted degree accounts for the strength of connection, or number of connections per pair of vertices (Wasserman and Faust, 1994). Degree is formulated as:

$$\deg(i) = \sum_{j=1}^{N} e_{ij} \tag{5.1}$$

where, e_{ij} is an edge connection between vertex i and vertex j. Weighted degree, on the other hand, is formulated as:

$$wdeg(i) = \sum_{j=1}^{N} e_{ij}^{w} \tag{5.2}$$

where, e_{ij}^{w} is a weighted edge between vertex i and vertex j.

Based on our model, the degree of a county represents the number of counties with which it shares common visitors. Weighted degree incorporates information on the total number of users in common. In the example shown in Figure 5.1, the degree of County A is 4 and the weighted degree is 160 (20 + 100 + 5 + 35). If a county has a relatively large degree or weighted degree then that means that its residents and/or visitors have also been present in many other counties. In other words, the county can be characterized as being highly mobile. Moreover the measures tell us how similar counties are in relation to one another in terms of the individuals who visit locations within them.

While degree and weighted degree provide information on the extent to which a county is connected to other counties via the location choices of individuals, they do not capture the distributional properties of the connections. For example, is a county equally connected to other counties or are the connections strong for only a subset of the counties? To capture this we use Shannon entropy, which is formulated as:

$$Entropy(i) = -\sum_{j=1}^{N} (p_{ij} \log p_{ij}) \tag{5.3}$$

where p_{ij} is the number of users county i has in common with j in relation to the total number of users in common to all other counties. For each county, the probabilities p_{ij} sum to one. We also use normalized entropy, which controls for the maximum entropy possible given the total number of other counties to which a county connects. Normalized entropy is bound by zero and one, where a value closer to one would reflect greater diversity than a value nearer to zero.

To capture the indirect connectedness of a county, we use vertex betweenness centrality. For an undirected and unweighted network, the betweenness centrality of a vertex represents the total number of shortest paths through the network that contain that vertex (Wasserman and Faust, 1994). Betweenness centrality is formulated as:

$$Bet(i) = \frac{g_{jk}(i)}{g_{jk}} \tag{5.4}$$

where g_{jk} is the total number of binary shortest paths in the network and g_{jk} (i) is the number of shortest paths that go through vertex i. In our model, the betweenness centrality of a county captures specifically how central the county is in relation to all other counties. In this regard, one can view it as an index of accessibility.

5.4 BRIGHTKITE LOCATION-SHARING SERVICES DATA

We now apply our model to a sample of Brightkite location-sharing services data, which was collected by the Stanford Large Network Database (SNAP) group (snap.stanford.edu/data/). The complete dataset contains 4491 143 check-ins made by Brightkite users over the period April 2008 to October 2010. Each record includes the latitude and longitude coordinates associated with the check-in, a time stamp for the check-in and a unique user identification (ID). We also utilize information on the Brightkite friendship network. The full undirected graph was comprised of 58228 individuals and 214078 edges (or ties between individuals) (Cho et al., 2011).

For practical purposes, we extract only those check-ins made to locations within the lower 48 states in the USA. Figure 5.2 shows the locations of these check-ins and how they were distributed across the counties in the lower 48 states in the USA. There were 2263843 check-ins in total. Approximately 95 per cent of these were made within the boundary of a metropolitan statistical area (MSA). Since check-ins tend to be

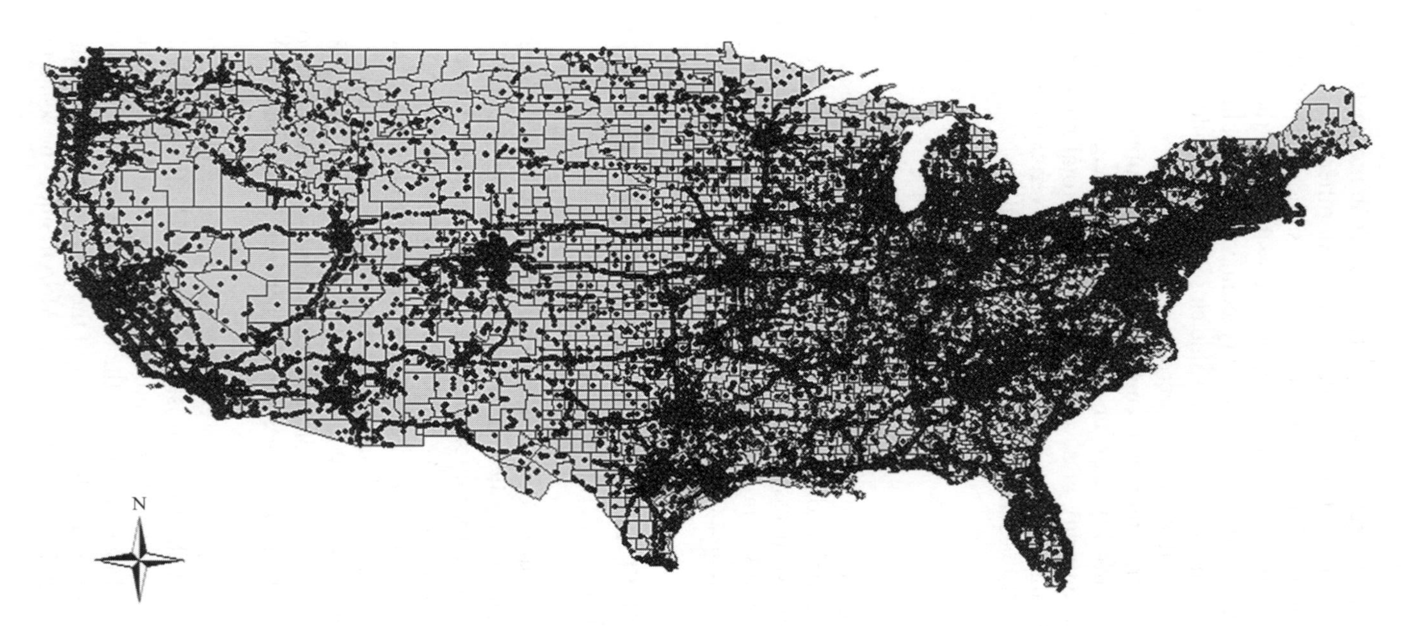

Figure 5.2 'Check-ins' made by Brightkite users in lower 48 states between April 2008 and October 2010

made on mobile devices, we expected to find that their locations would correspond to transportation facilities and systems. However we found that only 23 000 check-ins occurred within 0.05 miles of an Interstate highway. Moreover only 6486 were made at a primary airport, 2792 at an Amtrak station and 6176 at an intermodal passenger transportation facility.

Rather than using the latitude and longitude coordinates of the check-ins to designate their location, we used US counties for the geographic units. We first assigned check-ins to counties and then from each county extracted only the unique set of users. In other words, if an individual checked in more than once to a particular county we counted that individual only once. This information was then used to generate the two-mode network. This network was comprised of 21 001 unique Brightkite users and 2706 counties. Table 5.1 reports summary statistics for the 25 counties that had the most check-ins.

One can see that there is generally a positive association between check-in activity and population: that is, more populated areas tend to have more check-ins. However, some counties had more check-ins than what would be expected given their population levels: for example, Boulder (CO), Clark (NV) and the District of Columbia. For the most part these counties represent tourist destinations. From a practical standpoint, the counts also demonstrate that we have an adequate number of data points for each county from which to safely draw conclusions about the displacement properties and mobility patterns of the counties.

Table 5.2 reports the values of the mobility indicators that were introduced in the previous section. There are a couple of conclusions that can be drawn from the table. First, there is generally a positive association between the centrality of the counties and their check-in activity. Specifically, counties with more check-ins have higher degrees of displacement than those with fewer check-ins. Interestingly, Cook County, IL (Chicago Metropolitan Statistical Area) ranks exceptionally high for all the centrality indicators. Other counties with relatively high levels of centrality include those that are destinations or have international airports. Second, all of the normalized entropies are relatively close to one, which indicates that the distribution of common visitors for each county is fairly evenly distributed.

To assess whether or not there is a spatial association between the indicators, we generated Moran's I statistics and local indicators of spatial association (LISA) cluster maps for the following indicators: weighted degree, betweenness and normalized Shannon entropy (Figures 5.3 to 5.5). In each case, we used Queen contiguity to designate adjacency of the

Table 5.1 Summary statistics for counties with top 25 most check-ins

County	State	Population	Rank (top 100 populated counties)	Check-ins	Check-ins per population (100 000)	Total users	Check-ins per user
Los Angeles	California	9 889 056	1	133 755	1353	2417	55.33926
Denver	Colorado	619 968	a	75 060	12 107	1494	50.24096
Maricopa	Arizona	3 880 244	4	69 921	1802	1043	67.03835
King	Washington	1 969 722	14	69 542	3531	1330	52.28722
San Francisco	California	812 826	67	62 101	7640	2383	26.06001
Cook	Illinois	5 217 080	2	52 438	1005	1684	31.13895
New York	New York	1 601 948	20	49 515	3091	2135	23.19204
Santa Clara	California	1 809 378	16	43 801	2421	1412	31.02054
Boulder	Colorado	294 567	0	37 994	12 898	710	53.51268
Orange	California	3 055 745	6	33 595	1099	1029	32.6482
Fulton	Georgia	949 599	46	32 381	3410	896	36.13951
District of Columbia	District of Columbia	632 323	0 b	31 121	4922	1032	30.15601
Travis	Texas	1 063 130	38	30 231	2844	1061	28.49293
San Mateo	California	727 209	82	29 429	4047	1507	19.5282
Harris	Texas	4 180 894	3	29 279	700	718	40.77855
San Diego	California	3 140 069	5	27 797	885	905	30.71492
Jackson	Missouri	676 360	90	25 835	3820	480	53.82292
Allegheny	Pennsylvania	1 227 066	31	23 277	1897	343	67.86297
Dallas	Texas	2 416 014	9	23 198	960	735	31.5619
Multnomah	Oregon	748 031	77	22 160	2962	686	32.30321
Johnson	Kansas	552 991	a	21 874	3956	340	64.33529
Clark	Nevada	1 969 975	13	20 726	1052	1211	17.11478
Bexar	Texas	1756153	19	20 722	1180	406	51.03941
Tarrant	Texas	1 849 815	15	17 616	952	712	24.74157
Orange	Florida	1 169 107	34	17 482	1495	818	21.37164

Note: 'a' indicates that the county is not in the top 100 most populated; 'b': District of Columbia is not a county.

Table 5.2 Network statistics for counties with top 50 total check-ins ranked high to low

County	State	Degree	Weighted degree	Betweeness centrality	Entropy	Normalized entropy
Los Angeles	California	1883	23453	0.0107	2.6285	0.8026
Denver	Colorado	1975	21349	0.0147	2.7795	0.8438
Maricopa	Arizona	1785	14436	0.0080	2.8050	0.8641
King	Washington	1694	13948	0.0073	2.4074	0.8415
San Francisco	California	1789	23201	0.0093	2.5748	0.7917
Cook	Illinois	2131	24109	0.0209	2.8721	0.8781
New York	New York	1850	22890	0.0103	2.6006	0.8095
Santa Clara	California	1432	14756	0.0051	2.6054	0.8271
Boulder	Colorado	1171	8808	0.0020	2.4824	0.8095
Orange	California	1488	11311	0.0039	2.6723	0.8442
Fulton	Georgia	1731	12909	0.0088	2.7437	0.8624
District of Columbia	District of Columbia	1773	16557	0.0085	2.7131	0.8369
Travis	Texas	1683	13937	0.0090	2.4383	0.8510
San Mateo	California	1619	19139	0.0066	2.5897	0.8071
Harris	Texas	1757	11314	0.0073	2.5105	0.8681
San Diego	California	1470	10771	0.0042	2.7093	0.8592
Jackson	Missouri	1624	8168	0.0056	2.7714	0.8799
Allegheny	Pennsylvania	1539	6661	0.0029	2.5054	0.9140
Dallas	Texas	1752	11425	0.0081	2.4883	0.8754
Multnomah	Oregon	1501	9691	0.0034	2.3792	0.8551
Johnson	Kansas	1226	4919	0.0027	2.6675	0.8890
Clark	Nevada	1766	17031	0.0099	2.7789	0.8724
Bexar	Texas	1441	6888	0.0044	2.4933	0.9189
Tarrant	Texas	1862	13362	0.0104	2.5031	0.8653
Orange	Florida	1793	12611	0.0095	2.8276	0.8857

counties.[1] On the maps, the darkest grey indicates low-to-low positive spatial association, the next darkest grey represents high-to-high positive association, the lightest grey refers to negative association and white corresponds to insignificant association.

From the maps, it can be seen that all three indicators exhibit some degree of positive spatial association. Specifically, counties with high indices tend to be surrounded by others with high indices. The Moran's I statistics for weighted degree, betweenness and normalized Shannon entropy were 0.42951, 0.30 and 0.37699, respectively. All are significant at $p = 0.01$. The clusters of high-mobility counties are generally located in densely populated corridors, for example the West Coast and the Northeast.

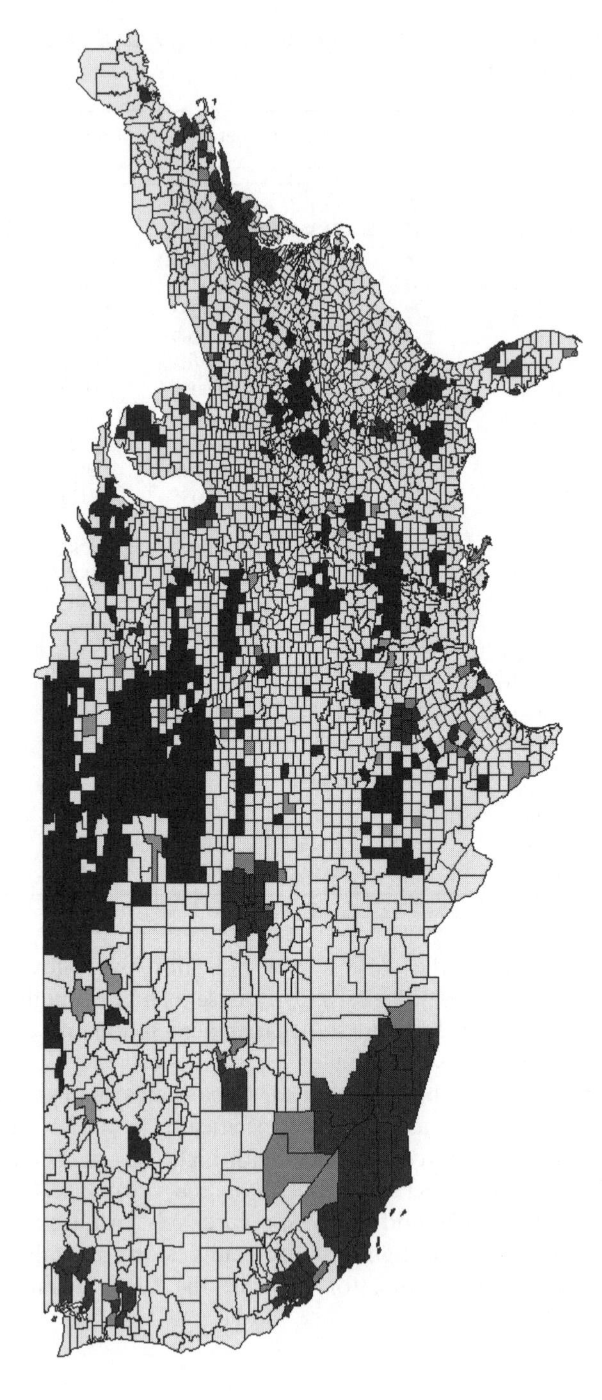

Figure 5.3 LISA cluster map of weighted degree for all counties in lower 48 states

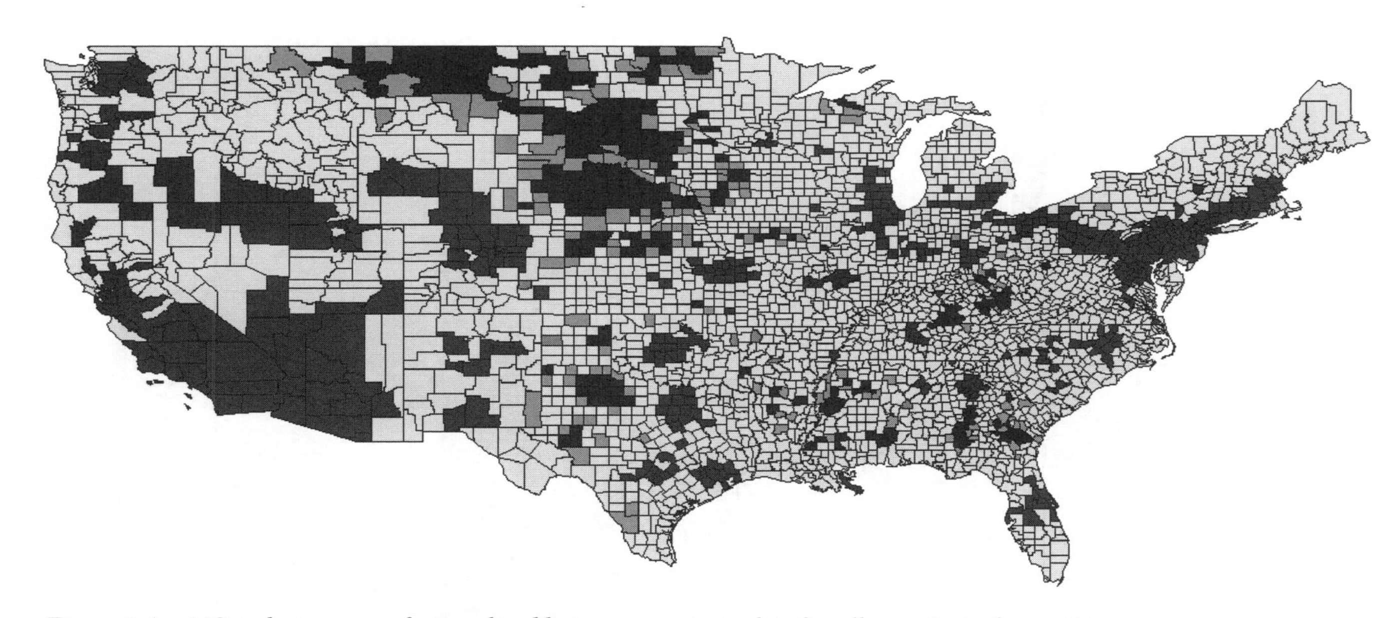

Figure 5.4 LISA cluster map of normalized betweenness centrality for all counties in lower 48 states

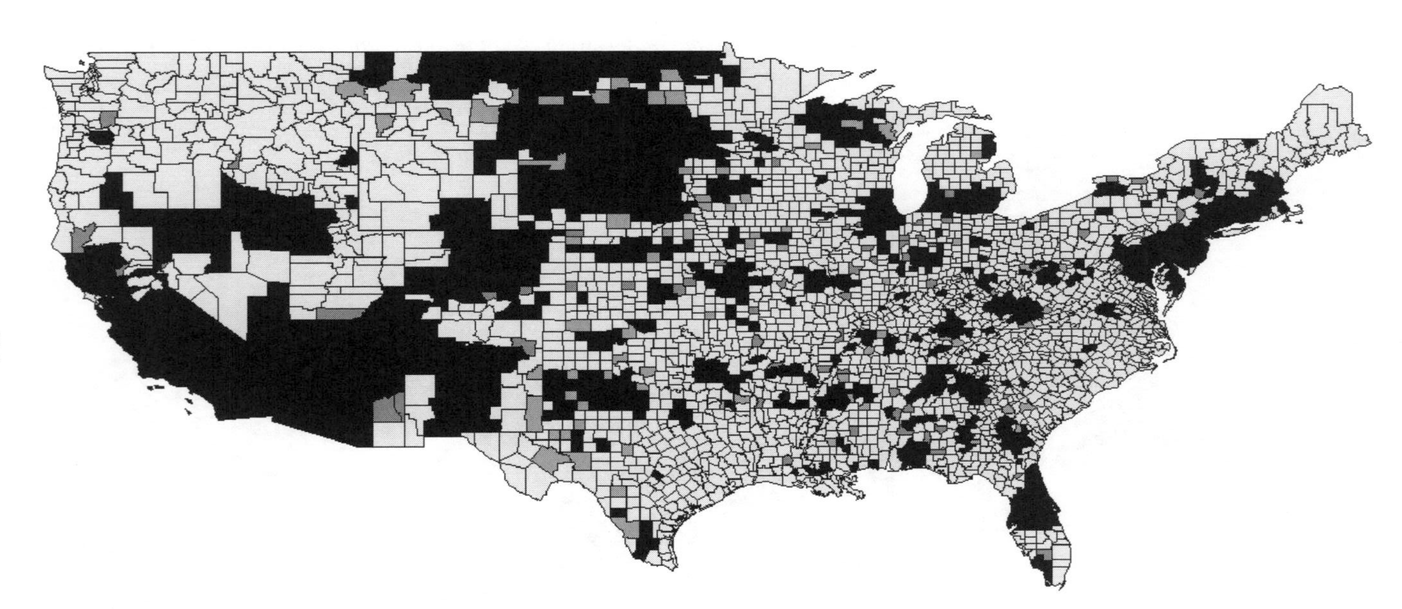

Figure 5.5 *LISA cluster map of entropy for all counties in lower 48 states*

5.5 REGIONAL VERSUS LONG-DISTANCE TRAVEL

The indicators just discussed measure only the level of mobility. They do not provide any information on the geographic range of movement for individuals who check into locations within the counties. To illustrate this, let us consider two counties with identical degree centrality. It could be the case that for one of them the counties with which it shares visitors are geographically close, while for the other they are more distant in space. To capture this, we formulate two additional indicators. One of these is formulated as follows:

$$RI(i) = \frac{wdeg(i)^*}{wdeg(i)} \tag{5.5}$$

where $wdeg(i)$ is the weighted degree for county i and $wdeg(i)^*$ is the weighted degree for only those edge connections within a certain distance threshold. The index is bounded by zero and one, where a value close to one reflects a minimal geographic range of movement, that is, high degree of regional rather than national mobility, and a value close to zero represents a greater geographic extent of movement.

Since the weighted degree of a county is a function of number of users who have checked in to that county as well as number of users tied to all other counties it has users in common with, we formulate an index that uses normalized edges as weights. Edge weights are normalized as follows:

$$e_{ij}^{Norm} = \frac{e_{ij}^{w}}{\sqrt{l_{ii}} * \sqrt{l_{jj}}} \tag{5.6}$$

where e_{ij}^{w} is the number of users county i and j have in common, l_{ii} is the number of users associated with county i, and l_{jj} is the number of users tied to county j. This measures how similar county i is to j in terms of common users, controlling for the total number of visitors to both county i and j. The normalized edge weight is bound by zero and one, where a zero would reflect no association and one perfect association. The normalized regional index, which we denote as NRI, is formulated as follows:

$$NRI(i) = \frac{wndeg(i)^*}{wndeg(i)} \tag{5.7}$$

where:

$$wndeg(i) = \sum_{j=1}^{N} e_{ij}^{Norm} \tag{5.8}$$

and $wndeg(i)^*$ is the weighted normalized degree for county i for only those counties within the distance threshold. Like equation (5.5), the index is bound by zero and one, with a value close to one indicating a high degree of regional rather than national movement and vice versa for a value close to zero. For the distance threshold, we used 100 km. This particular distance was used to ensure that each county had at least one other county within the threshold.

From Table 5.3, which reports the indices, one can see that all of the counties have relatively high ranges of movement – that is, there are a large number of national trips in relation to regional trips. Counties with the least regional range of movement – for example, Orange County, FL (Orlando MSA) and Clark County, NV (Las Vegas) – tend to be tourist

Table 5.3 Indicators of regional mobility for 25 counties with the most check-ins

County	State	Regional index (RI)	Normalized regional index (NRI)
Los Angeles	California	0.0389	0.0086
Denver	Colorado	0.1615	0.0670
Maricopa	Arizona	0.0221	0.0131
King	Washington	0.0596	0.0499
San Francisco	California	0.1735	0.0622
Cook	Illinois	0.0524	0.0487
New York	New York	0.2079	0.1351
Santa Clara	California	0.2123	0.0918
Boulder	Colorado	0.3017	0.1703
Orange	California	0.0985	0.0133
Fulton	Georgia	0.1407	0.1683
District of Columbia	District of Columbia	0.2175	0.1653
Travis	Texas	0.0667	0.1016
San Mateo	California	0.1865	0.0686
Harris	Texas	0.0416	0.0610
San Diego	California	0.0485	0.0164
Jackson	Missouri	0.1871	0.1472
Allegheny	Pennsylvania	0.0754	0.1453
Dallas	Texas	0.1175	0.1033
Multnomah	Oregon	0.0860	0.0878
Johnson	Kansas	0.2680	0.1966
Clark	Nevada	0.0036	0.0078
Bexar	Texas	0.0922	0.1234
Tarrant	Texas	0.0793	0.0712
Orange	Florida	0.0743	0.0726

destinations. With a few exceptions, the counties with the highest degree of national trips in relation to regional trips are located in the western and southern portions of the country.

5.6 MOBILITY SHEDS

We now examine whether or not we can identify distinct regions in the US defined by the mobility patterns of individuals. More specifically, we attempt to detect 'mobility sheds', which comprise individuals who tend to travel within the same geographic range. We use a community detection algorithm to extract these mobility sheds.

A community detection algorithm attempts to segment a network into clusters of densely connected nodes or edges. There are many different types of community detection algorithms (Fortunato, 2009). For the purposes of this study, we use Blondel's modularity optimization algorithm, which has a number of appealing properties. First, it yields information on the hierarchical community structure of a network. Second, the algorithm is very efficient and can be applied to large-scale networks. Lastly, it has been found to produce higher-quality communities than other algorithms (Blondel et al., 2008).

Blondel community detection is based on optimization of a modularity function. For an unweighted network, this function is:

$$Q = \frac{1}{2m}\sum_{i,j}\left[e_{ij} - \frac{d_i d_j}{2m}\right]\delta_{c_i c_j} \tag{5.9}$$

where e_{ij} is the edge between node i and j, $d_i = \sum_j e_{ij}$, $d_j = \sum_j e_{ji}$, $m = \frac{1}{2}\sum_{ij} e_{ij}$, c_i is the community node i is assigned to, and c_j the community node j belongs to. The Kronecker delta $\delta\,(u, v)$ is equal to one if $u = v$ and zero otherwise. This function measures the density of edges within communities in relation to those between communities. It has a lower bound of -1 and upper bound of 1, where a value closer to one characterizes a more well-defined, sophisticated community structure (Blondel et al., 2008).

The one-mode network was too dense to detect a distinct community structure. Therefore, we applied the community detection algorithm instead to the original bipartite network. These communities are shown in Figure 5.6.

From this, it is apparent that there are distinct regions in the US based on the common mobility patterns of individuals. These communities, which we refer to as mobility sheds, represent regions in which the same sets of people are travelling. The communities detected are generally

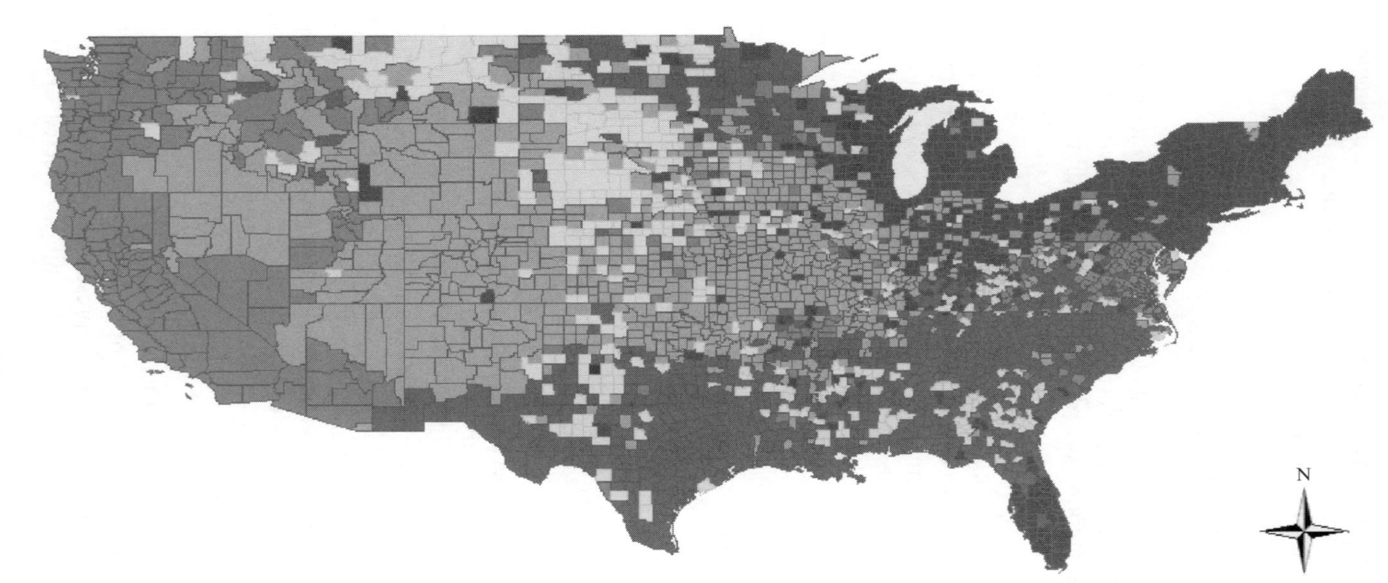

Note: the different shades of grey correspond to the different communities detected from the bipartite network.

Figure 5.6 Mobility sheds based on Blondel community detection on the 'socio-spatio' bipartite network

comprised of contiguous counties. However, there are some spatial outliers or counties that fall outside their respective mobility sheds. These counties perhaps represent locations with relatively strong long-distance ties. Also, relatively small communities are located in areas of the US which contain the most densely populated counties, that is, the Northeast, the Mid-Atlantic and the upper Midwest. This is also consistent with what Noulas et al. (2011a) found in their study.

5.7 CONCLUDING REMARKS

The analysis provides some insight into what types of trips are represented by location-sharing services data. The Brightkite data captures both short- and long-distance trips. Moreover many of the check-ins occurred in counties considered as tourist or vacation destinations, suggesting that location-sharing service data may be a proxy for leisure travel.

We found that there are regional variations in all of the indicators of mobility that were developed in this chapter. Additionally, there are distinct regions defined by the mobility patterns of the Brightkite users. Further research could tie these patterns to underlying socio-economic characteristics and other locational attributes to gain a richer explanation for the regional variations that were discovered.

The results of our analysis have implications for information diffusion and epidemiological modelling. In particular, counties with the highest degree of social and spatial centrality may be locations from which information or disease could spread most rapidly to all other counties. Further research could examine this, bringing in temporal information on the check-ins and/or using more geographically fine-grained specifications for the locations.

NOTE

1. According to the Queen rule of adjacency, a location is contiguous with another location if it shares a common border. Unlike the Rook rule, the Queen rule considers all locations and not just those that are located immediately to the east, west, north and south.

REFERENCES

Blondel, V.D., J.L. Guillaume, R. Lambiotte and E. Lefebvre (2008), 'Fast unfolding of communities in large networks', arXiv.org Physics, arXiv: 0803.0476.

Candia, J., M. Gonzalez, P. Wang, T. Schoenharl, G. Madey and A.L. Barabasi (2008), 'Uncovering individual and collective human dynamics from mobile records', *Journal of Physics A: Mathematical and Theoretical*, **41**, 1–11.

Cheng, Z., J. Caverlee and K. Lee (2011), 'Exploring millions of footprints in location sharing services', Association for the Advancement of Artificial Intelligence.

Cho, E., S.A. Myers and J. Leskovec (2011), 'Friendship and mobility: user movement in location-based social networks', ACM SIGKDD International Conference on Knowledge Discovery and Data Mining (KDD).

Davis, A., B.B. Gardner and M.R. Gardner (1941), *Deep South*, Chicago, IL: University of Chicago Press.

Elwood, S., F.G. Goodchild and D.Z. Sui (2012), 'Researching volunteered geographic information: spatial data, geographic research, and new social practice', *Annals of the Association of American Geographers*, **102** (3), 571–590.

Fortunato, S. (2009), 'Community detection in graphs', arXiv: 0906.0612 [physics.soc-ph].

Freeman, L. and V. Duqueene (1993), 'A note on colorings of two mode data', *Social Networks*, **15**, 437–441.

Gonzalez, M., C. Hidalgo and A.L. Barabasi (2008), 'Understanding individual human mobility patterns', *Nature Letters*, **5**, 779–782.

Greene, D. and P. Cunningham (2010), 'Spectral co-clustering for dynamic bipartite graphs', Technical Report UCD-CSI-2010–05, School of Computer Science and Informatics, University College Dublin.

Lawlor, A., C. Coffey, R. McGrath and A. Pozdnoukhov (2012), 'Stratification structure of urban habitats', Working Paper, National Centre for Geocomputation, National University of Ireland Maynooth.

Noulas, A., S. Scellato, S. Lambiotte, M. Pontil and C. Mascolo (2011a), 'A tale of many cities: universal patterns in human urban mobility', PLOS ONE.

Noulas, A., S. Scellato, C. Mascolo and M. Pontil (2011b), 'An empirical study of geographic user activity patterns in Foursquare', Association for the Advancement of Artificial Intelligence, www.aaai.org.

Scellato, S. and C. Mascolo (2011), 'Measuring user activity on an online location-based social network', IEEE.

Schmutte, I. (2014), 'Free to move? A network analytic approach for learning the limits of job mobility', *Labour Economics*, **29**, 49–61.

Song, C., Z. Qu, N. Blumm and A.L. Barabási (2010), 'Limits of predictability in human mobility', *Science*, **327**, 1018–1021.

Wasserman, S. and K. Faust (1994), *Social network analysis: methods and applications*, Cambridge: Cambridge University Press.

Zickhur, K. (2012), 'Three-quarters of smartphone owners use location-based services', Pew Research Center's Internet and American Life Project, http://pewinternet.org/Reports/2012/Location-based-services.asp.

PART II

The social and spatial dimension of accessibility

6. Spatial organization and accessibility: a study of US counties

Andrea De Montis, Simone Caschili and Daniele Trogu

6.1 INTRODUCTION

Commuting is a phenomenon that widely affects contemporary societies confronted with the opportunity to work in places that are sometimes located at a considerable distance from residential districts. The areas in which we live are organized to allow for the daily movement of workers and students, calling for efficient transportation systems in order to minimize costs. In regional science, accessibility is often used to evaluate, among other characteristics, the effectiveness and quality of transportation systems. This multifaceted concept encapsulates the ability of a certain category of people to reach a given location and clearly depends on the logistics of transportation infrastructures. In the literature accessibility is assessed by means of several indicators, whose functional structure is generally related to two factors: job opportunities and the transport costs that people sustain to reach a location. Whatever accessibility measure is applied, the level of accessibility often depends on spatial location.

In this chapter we apply spatial autocorrelation analyses to investigate the spatial distribution of commuters' accessibility across US counties. Commuting is a mushrooming phenomenon in the US: approximately 25 per cent of the total US labour force (about 32 million people) commute from a distance radius of 25 minutes, according to the dataset on commuting behaviour provided by the US Census Bureau (www.census.gov). This information refers to commuting trips in the counties of the 50 states and the District of Columbia (DC). We consider inbound and outbound movements to and from each of the 3141 counties on the US mainland.

In section 6.2 we present a brief review of state-of-the-art applications on spatial autocorrelation for accessibility and commuting, followed in section 6.3 by a discussion on the foundations of spatial autocorrelation methods. In section 6.4 we discuss accessibility and network characteristics that are of interest for our study. Section 6.5 is dedicated to the presentation of results concerning spatial autocorrelation analysis of accessibility and macro-socio-economic variables. In the concluding section we provide a discussion and interpretation of the results and suggest further avenues for research.

6.2 ACCESSIBILITY AND SPATIAL STATISTICS FOR URBAN AND TRANSPORT STUDIES: STATE OF THE ART

In this section, we provide a brief review of state-of-the-art studies concerning two issues: accessibility, and spatial autocorrelation tools for investigating accessibility, commuting and labour patterns. Accessibility is a crucial concept in transport analysis and planning. The widespread adoption of this concept stems from the work of Hansen (1959) and Weibull (1976), who first formulated a quantitative definition. The main idea underlying their studies is that the accessibility associated with places can be measured as a potential for opportunities set against the cost paid by commuters to move through space and time. Many reviews discuss this concept and report on a range of methods and indicators adopted in accessibility modelling (see *inter alia* Baradaran and Ramjerdi, 2001; De Montis and Reggiani, 2012, 2013; Geurs and van Wee, 2004; Handy and Niemeier, 1997; Jones, 1981; Martín and Reggiani, 2007; Wu and Hine, 2003).

In the field of transportation, travel behaviour and urban analysis, Páez and Scott (2004) propose an extensive review of spatial statistics methods and tools (including spatial association, spatially autoregressive models and local statistics of spatial association). A number of applications have considered labour markets, the spatial distribution of (un) employment (Molho, 1995; McMillen, 2004; Patacchini and Zenou, 2007), and housing prices (Gillen, 2001; Fingleton, 2006).

Our interest in this chapter focuses on the spatial patterns of commuting. Many authors have applied spatial autocorrelation analysis to ascertain the geographical dependence of commuters' behaviour. Griffith (2007) studies commuting in Germany at NUTS 3[1] level (districts; *Regierungsbezirk* in German) and finds a crucial interplay between spatial interaction modelling and spatial autocorrelation analysis. His arguments confirm an earlier hypothesis (Curry, 1972) that 'spatial autocorrelation effects are confounded with distance decay effects during the estimation of simple gravity model parameters' (Griffith, 2007, p. 38). Vandenbulcke et al. (2011) study the spatial pattern of bicycle commuting in Belgium, which is exemplified by two main clusters of bicycle commuting municipalities in the northern Flemish and southern Walloon regions. In the first case bicycle commuting is positively influenced by neighbouring municipalities, while in the southern Walloon region there is a negative spatial influence. Wang (2001) studies the distortive effects of positive spatial autocorrelation on intra-urban data in order to analyse the intra-urban variations of average commuting time and distance in Columbus, United States. Kawabata and Shen (2007) develop spatial analyses to investigate the association between job accessibility and commuting time for public transit and private cars within the San Francisco Bay area in the US. They take into account spatial autocorrelation, spatial lag and spatial error models using spatial regression models.

In this section we have briefly examined some research studies that are of interest for our discussion. Although spatial autocorrelation has been extensively utilized in the study of commuting, to the best of our knowledge no study has employed this technique in order to understand the dependency of accessibility in regional units on spatial and socio-economic variables. In the next section we discuss the foundations of the spatial autocorrelation statistics that are of interest for this study.

6.3 SPATIAL AUTOCORRELATION: UNIVARIATE AND BIVARIATE ANALYSIS

Spatial autocorrelation analysis is a statistical methodology that takes into account the spatial relationships (adjacency, contiguity, proximity, and so on) between locations, where a phenomenon occurs. The first law of geography by Tobler (1970) asserts that 'Everything is related to everything else, but near things are more related than distant things', which is indicative of the basic concepts supporting spatial autocorrelation. This concept is relevant, as most statistical analyses are based on the assumption that in each sample observations are independent of each other (Getis, 2007). However, exogenous variables are sometimes dependent; this occurs, in particular, for spatial phenomena. In this chapter the term 'spatial phenomenon' is used to identify a phenomenon that we observe in locations within a given spatial domain. Spatial characteristics may not always be randomly distributed but have systematic patterns over the same domain. This is the case of spatial dependence. Spatial autocorrelation measures the strength of spatial dependence and tests for the spatial independence of observations. Spatial autocorrelation is positive when variables have similar trends in neighbouring locations and negative when variables have dissimilar trends in close spatial association.

It is very important to detect spatial autocorrelation, as its underestimation could generate errors in modelling forecasts. In fact, violation of the condition of independence modifies the error specification of models and can lead to incorrect forecasts, especially when modelling refers to the prediction of spatial behaviour. Spatial autocorrelation analysis thus acts as a diagnostic tool for inferential statistics models. In this case, the detection of spatial autocorrelation leads to the introduction of a spatial relationship between the variables of a model. In this way, it is also possible to represent missing variables for a specific location, using a spatial regression model, for example (Cliff and Ord, 1969). According to Griffith (2009), the term 'correlation' implies redundant information: if two variables are perfectly correlated, it is possible to know the value of one variable through the value of the other. Spatial autocorrelation extends the concept of correlation to geo-referenced data: the value of a variable is associated with the value of the same variable in another (nearby) location.

Under a quantitative perspective, Moran (1950) proposed a measure of spatial autocorrelation that obeys the following expression:

$$I = \frac{\left(n \sum_{i} \sum_{j} w_{ij}(x_i - \mu)(x_j - \mu)\right)}{\left(\sum_{i} \sum_{j} w_{ij} \sum_{i} (x_i - \mu)^2\right)} \tag{6.1}$$

In equation (6.1), μ represents the mean of the variable x in n observations, x_i and x_j are the observations in location i and j, and w_{ij} represents the so-called 'spatial weight'. The spatial weight w_{ij} is a term that takes into account spatial interconnectedness (that is, number of neighbours, length of shared boundary or distance between locations). Moran's index ranges from -1 to $+1$: values close to $+1$ represent a strong positive spatial autocorrelation (similar closely located trends), while values close to -1 indicate negative autocorrelations (dissimilar closely located trends).

Moran's index measures spatial autocorrelation across a set of spatial units (global approach). There are significant limitations to the identification of local spatial units when spatial autocorrelation is more significant. Anselin (1995) proposed a measure called the local indicator of spatial association (LISA) in order to overcome this limitation. According to Anselin, any statistical measurement can be considered a LISA, if it satisfies two requirements: (1) each observation conveys an indication of significant spatial clustering in its neighbourhood; (2) the summation of all observations is proportional to a global indicator of spatial association. The LISA provides spatial clusters where autocorrelation occurs with respect to certain significance values (Anselin, 1995).

Given its characteristics, spatial autocorrelation is often used to study the diffusion in space of phenomena which are significantly related to geographic location (the diffusion of diseases, market trends, and so on). In this chapter our interest revolves around accessibility for commuters in the counties of the US. We want to check whether accessibility is spatially self-correlated or correlated to relevant socio-economic characteristics.

6.4 ACCESSIBILITY AND COMMUTING NETWORK FEATURES

Since the Second World War, the number of people who commute to work has continuously increased. Based on recent estimates from the 2009 Census, about 91 per cent of US workers travel to work by private (86 per cent) or public means (5 per cent), while only 4 per cent work at home and 5 per cent travel to work by other means (walking, by bike, taxi, and so on).[2] Similar figures are observed in the European Union (EU), where 84 per cent of inland commuters use cars; 7 per cent use railways, trams and metros; and 8 per cent use buses and coaches.[3] Commuting is thus an important phenomenon that affects our daily lives and life style. Commuting is more than travelling from home to work: it consumes time and generates both monetary costs and social costs (pollution, stress and so on). In the US people spend, on average, 52 minutes per day commuting, while in Europe they spend an average of about 45 minutes. According to the basic economic idea of compensation (Horner, 2008), people make the decision to commute or not by trying to optimize the balance between travel time and other benefits such as higher salary or residing in a more attractive environment. The dichotomous choice between job opportunities and residential locations also shapes an area's landscape and land use.

Commuters generate a complex network of relationships in a territory when moving from home to work. In this sense it is interesting to study the patterns that emerge from examining commuting as networks (De Montis et al., 2007; De Montis et al., 2011; Lenormand et al., 2012). If we think of commuting systems as networks, a territory can be modelled as a set of nodes and links: nodes represent aggregated poles that can be either the origin or destination of commuting trips, while links encapsulate commuting relationships between two poles. Because a topological network does not take the magnitude of commuting between nodes into consideration, authors usually prefer to use weighted commuting networks. In this case, a weight is associated to each link as a function of the number of commuters that travel between two points. Several metrics can be derived from a network analysis, which helps one to understand the territorial patterns emerging from commuting (for a review, see De Montis et al., 2007).

The accessibility of an area is closely connected to the concepts of commuting, land use, jobs and housing location. Accessibility measures urban sprawl decentralization (Lucy and Phillips, 1997; Van Ham et al., 2001) and investigates the uneven distribution of economic activities (Reggiani, 2008). According to Horner (2008) accessibility can focus on: (1) a macro formulation that takes into consideration the location of zones in terms of proximity and potential interaction; and (2) a micro formulation which assesses the choices of individuals in finding activities proximal to them. In this chapter we consider a macro formulation of accessibility, based on spatial interaction models (for a review of these concepts, see Reggiani, 2008).

If we call A_i the accessibility of zone i and Ω the set of zones j that are interconnected to i, we can write the analytical macro formulation of accessibility as follows:

$$A_i = \sum_{j \in \Omega} D_j f(c_{ij}) \tag{6.2}$$

where D_j is a measure of activities in j; and $f(c_{ij})$ is the impedance function which reflects the factors deterring users from moving from i to j. According to Reggiani (2011), five different formulations of $f(c_{ij})$ have been applied in the literature:

- exponential decay function;
- exponential normal decay function;
- exponential square root decay function;
- log-normal decay function;
- power decay function.

The use of a specific decay function depends on the setting under scrutiny. In fact power-law decay better suits long-distance interactions (Fotheringham and O'Kelly, 1989), while exponential functions are more suitable for interactions in homogeneous areas (Wilson, 1971).

Caschili and De Montis (2013) have examined the accessibility of US counties, integrating a complex analysis of the US commuting network with the construction of accessibility indicators based on travel cost and a spatial interaction model (SIM). They found that SIM-based indicators provide a better representation of accessibility. In particular, an accessibility indicator with an exponential impedance function displays a strong positive correlation with network centrality. On the East Coast and in the Midwest, the population seems to be concentrated around large to mid-sized cities. As one travels further west, population and county accessibility levels decrease along with the number of larger cities. A qualitative and quantitative analysis suggests that accessibility increases with a rise in population, population density and per capita income, while weak positive correlation was found between the size of a county and its accessibility. Caschili and De Montis (2013) concluded that accessibility might be explained by a combination of several interconnected variables such as level of infrastructure, population, jobs, services and wealth.

While Caschili and De Montis (2013) applied a geographical analysis of accessibility determinants in US counties, they did not investigate the spatial auto-dependence of county accessibility. In order to fill this gap, in this chapter we apply spatial autocorrelation analyses to investigate the spatial pattern of accessibility in US counties. In the next section we implement univariate global and local spatial autocorrelation analyses of commuting accessibility indicators. In addition, bivariate global and local spatial autocorrelation analyses are applied to assess the statistical association between accessibility indicators and both residential population and average income.

6.5 RESULTS

The spatial autocorrelation analysis presented in this chapter is implemented using GeoDa software (Anselin, 2003). We first carry out a univariate analysis in order to study spatial correlation in the accessibility of US counties. We then implement a bivariate analysis in order to investigate the spatial relation between accessibility indicators and some sociodemographic measures. Our goal is to determine whether accessibility may be spatially conditioned by the socio-economic features of a territory.

6.5.1 Preliminary Analysis

The spatial distribution of accessibility was first determined by means of classical statistical tools, such as quantile analysis. We then mapped the results of this analysis in order to cluster US counties with a similar level of accessibility (quantile map, Figure 6.1). This preliminary analysis provided us with a rough idea of the emergence of spatial patterns that may be due to spatial autocorrelation. We scrutinized accessibility indicators using both the exponential decay function (Acc_{exp}) and the power decay function (Acc_{pwr}). There is a high correlation between the distributions of these two indices (coefficient of correlation = 0.78); for this reason and for ease of reading we will comment on the results of accessibility constructed

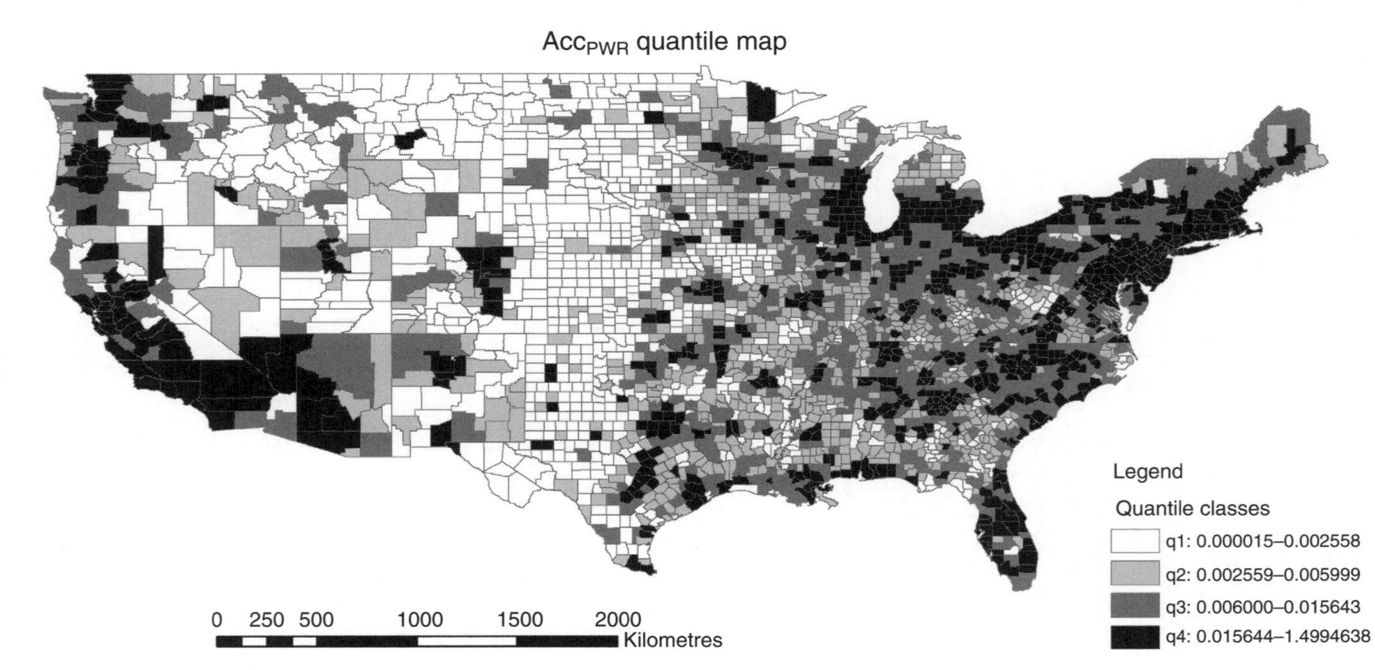

Figure 6.1 Quantile map of the accessibility indicator Acc_{pwr} in US counties

with the power decay function. We decided to use this approach because it has been pointed out by Caschili and De Montis (2013) and in other studies (Fotheringham and O'Kelly, 1989) that the power decay function is a better representation of the accessibility of US counties and long-distance interactions in general.

According to the quantile analysis (Figure 6.1), the least accessible counties (lowest accessibility values) are found in the middle of the US; while the most accessible counties are located around metropolitan areas, that is, clusters in the South West and North East coastal US.

6.5.2 Spatial Autocorrelation Analysis

Spatial autocorrelation analysis implemented for the accessibility indicator Acc_{pwr} has adopted both a global and a local approach and has been repeated with uni- and bivariate techniques. Before delving deeply into spatial autocorrelation analysis, it is worth clarifying some technical issues with respect to the calculation of the spatial weight matrix (SWM), which represents the level of adjacency between counties. In general terms, if i and j are two spatial units, the weight w_{ij} represents the so-called 'spatial influence' of unit j on unit i. There are various methods for evaluating SWM; in some cases, the weight may reflect the distance between centroids of units or the length of the shared boundaries (Cliff and Ord, 1969). In this study we evaluate SWM taking into account the first level of contiguity, so that each weight w_{ij} is equal to 1 if i is adjacent to j, and 0 otherwise. We use the Queen case to define the contiguity of counties; that is, if they share an administrative border.

The contiguity distribution of US counties is depicted in the histogram in Figure 6.2. The most populated classes include counties that present five, six and seven neighbours and these account for a significant share of the total distribution (75 per cent). According to the spatial autocorrelation methods illustrated by Anselin (2003), Acc_{pwr} has been standardized and afterwards analysed in equation (6.1).

6.5.3 Univariate Spatial Autocorrelation Analysis

Univariate global spatial autocorrelation analysis indicates a strong spatial dependence on the accessibility of counties. The value of Moran's index is 0.54. Since most points fall within the upper right quadrant and near the origin of the axes (Figure 6.3), there is definitely a positive spatial correlation in the accessibility of US counties. Counties with low accessibility generally have neighbours with similar accessibility. A low–low pattern emerges (see Figure 6.4).

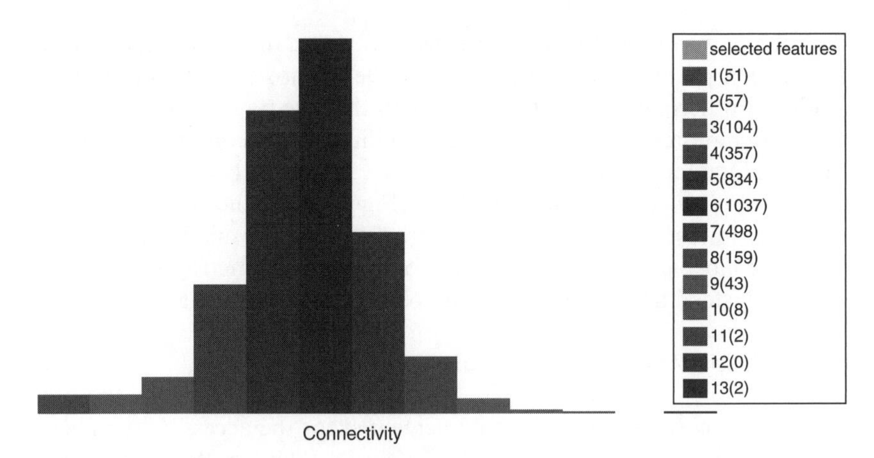

Figure 6.2 Contiguity analysis of US counties

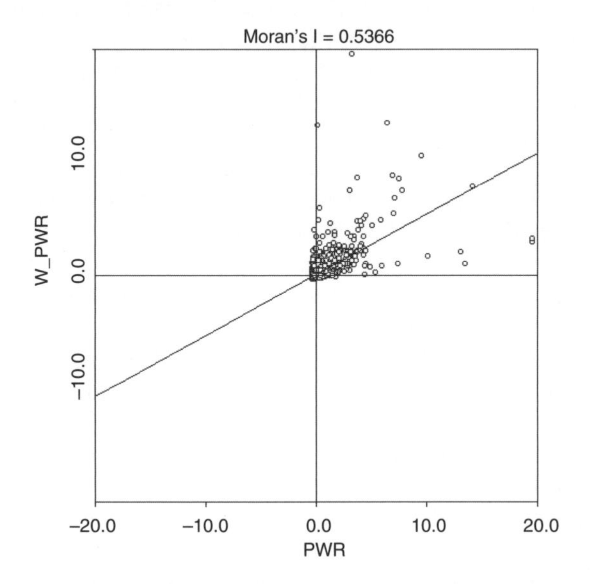

Figure 6.3 Univariate global spatial autocorrelation for Acc_{pwr}

However, spatial dependence may be even more significant in specific regions. We have therefore developed a finer inspection of spatial dependence and its significance through the local indicator of spatial autocorrelation (LISA). This methodology allows one to highlight homogeneous areas (that is, clusters of counties) where spatial dependence is stronger and more statistically significant.

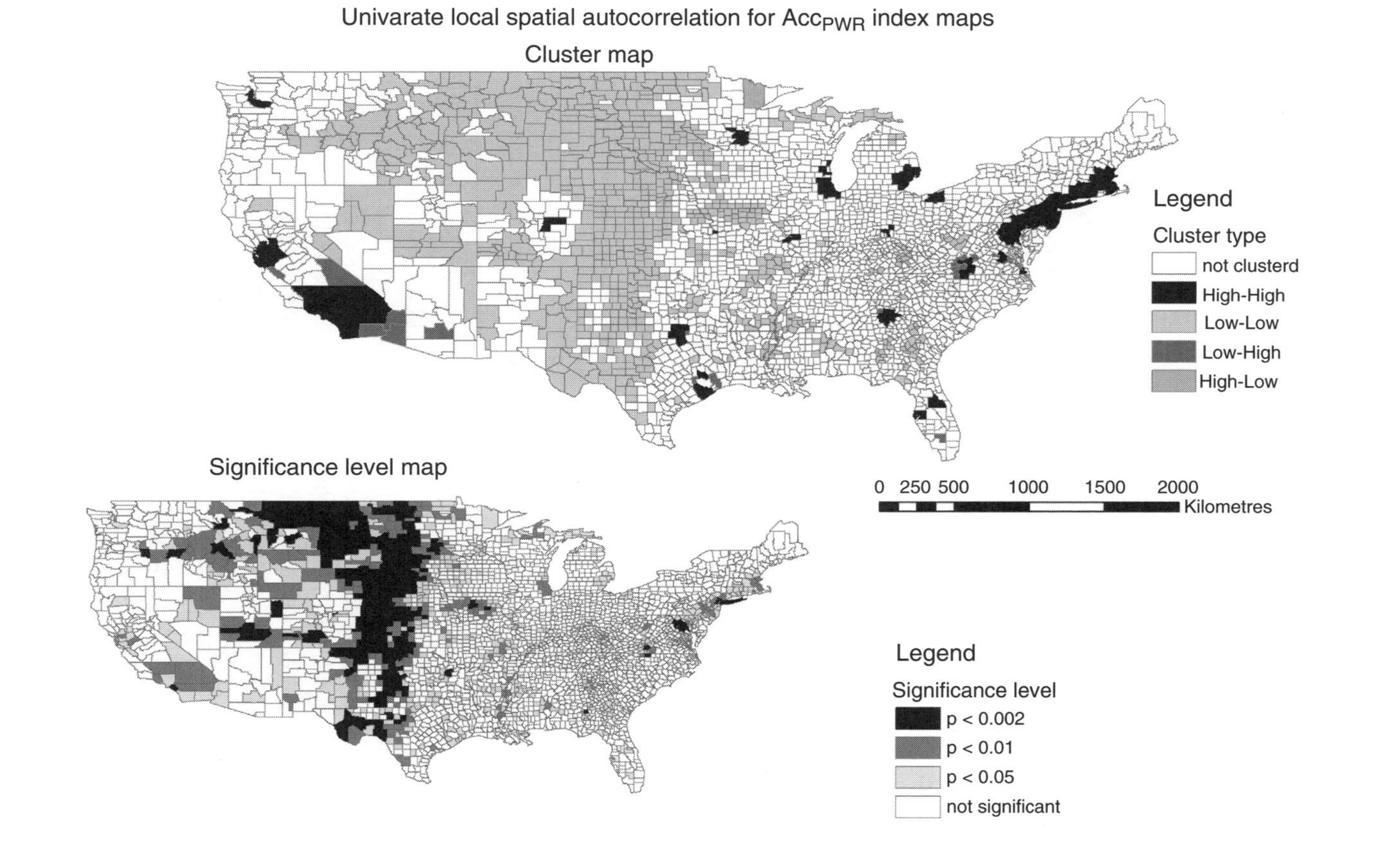

Figure 6.4 LISA cluster map for US counties for accessibility (top) and significance level map (bottom)

The results of the LISA and the significance map are reported in Figure 6.4. The cluster map classifies US counties according to association patterns, while the significance map reports the level of significance (p-value) of the corresponding spatial dependence attribute. In the cluster map, counties with high accessibility that are surrounded by highly accessible counties are visualized in black; lighter grey clusters represent less accessible counties surrounded by other counties with low accessibility. There is a tendency for the counties' accessibility to cluster into homogeneous areas. As discussed before, those counties fall into the first and third quadrant in Figure 6.3. Highly accessible counties surrounded by counties with low accessibility and low accessibility counties surrounded by highly accessible counties represent spatial outliers. This second group of counties exhibit negative spatial autocorrelation and they are represented by points in the second and fourth quadrants in Figure 6.3. As also indicated by Moran's index, the 'low–low cluster' is the class most often associated with high values of statistical significance (p-value < 0.002, coloured black in the significance level map).

We can thus see that a large cluster of positive low–low correlation emerges in the mid-US (Figure 6.4). This confirms a trend already detected through quantile analysis. Accessible counties clustered with other highly accessible counties are located on the northern Atlantic and southern Pacific coastlines and in the Great Lakes region. The significance level map indicates that autocorrelation is also statistically significant in these regions (p-values always lower than 0.05).

6.5.4 Bivariate Spatial Autocorrelation Analysis

Caschili and De Montis (2013) have investigated the statistical correlation between accessibility and two relevant socio-economic variables: population and income per capita. They have found that there is a strong positive correlation between accessibility and the number of residents in a county, while there is weak positive correlation between accessibility and average income. The scatter plot of bivariate analysis reported in Figure 6.5 indicates a weak positive spatial autocorrelation (Moran's index of 0.33), showing that, globally, population patterns are weakly correlated with accessibility.

Applying the LISA to scrutinize the spatial bivariate autocorrelation analysis of accessibility versus residential population we have detected similar clusters to the univariate analysis (Figure 6.6). Two pronounced clusters of counties, marked in black, emerge in the north-east and south-west, corresponding to regions where counties are highly accessi-

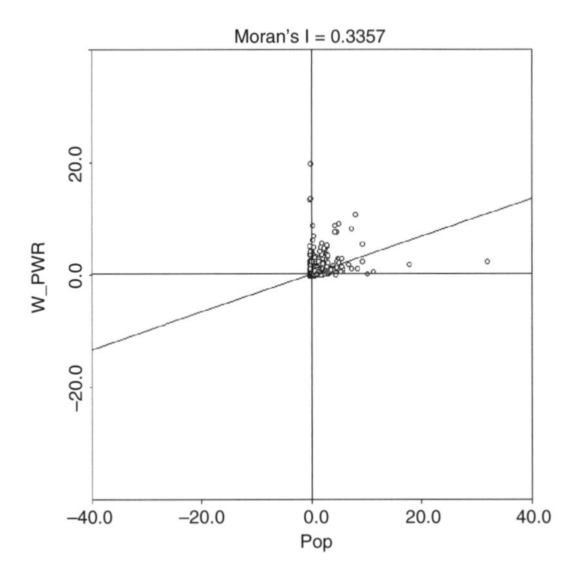

Figure 6.5 Spatial bivariate global autocorrelation analysis of accessibility versus population

ble and have a higher density of population. A large cluster of relatively less developed, rural counties, marked in lighter grey, is located in the Mid-West and includes units with low values of both accessibility and population.

A different picture emerges when we turn to bivariate spatial autocorrelation analysis of accessibility indicators and income per capita. As shown in Figure 6.7, global bivariate analysis yields a higher spatial correlation (Moran's index of 0.41).

In Figure 6.8, we display the results of the LISA for accessibility and income per capita. The cluster map is quite different from the bivariate analysis of accessibility and residential population. Apart from clusters already detected and marked in black (for high accessibility and high income per capita) and lighter grey (low–low), negative spatial autocorrelation is detected in mid-US counties (high–low) and southern Pacific coastal US (low–high). In these cases we find counties with a low accessibility which have neighbours with high average income, and counties with a high accessibility which have neighbours with low average income.

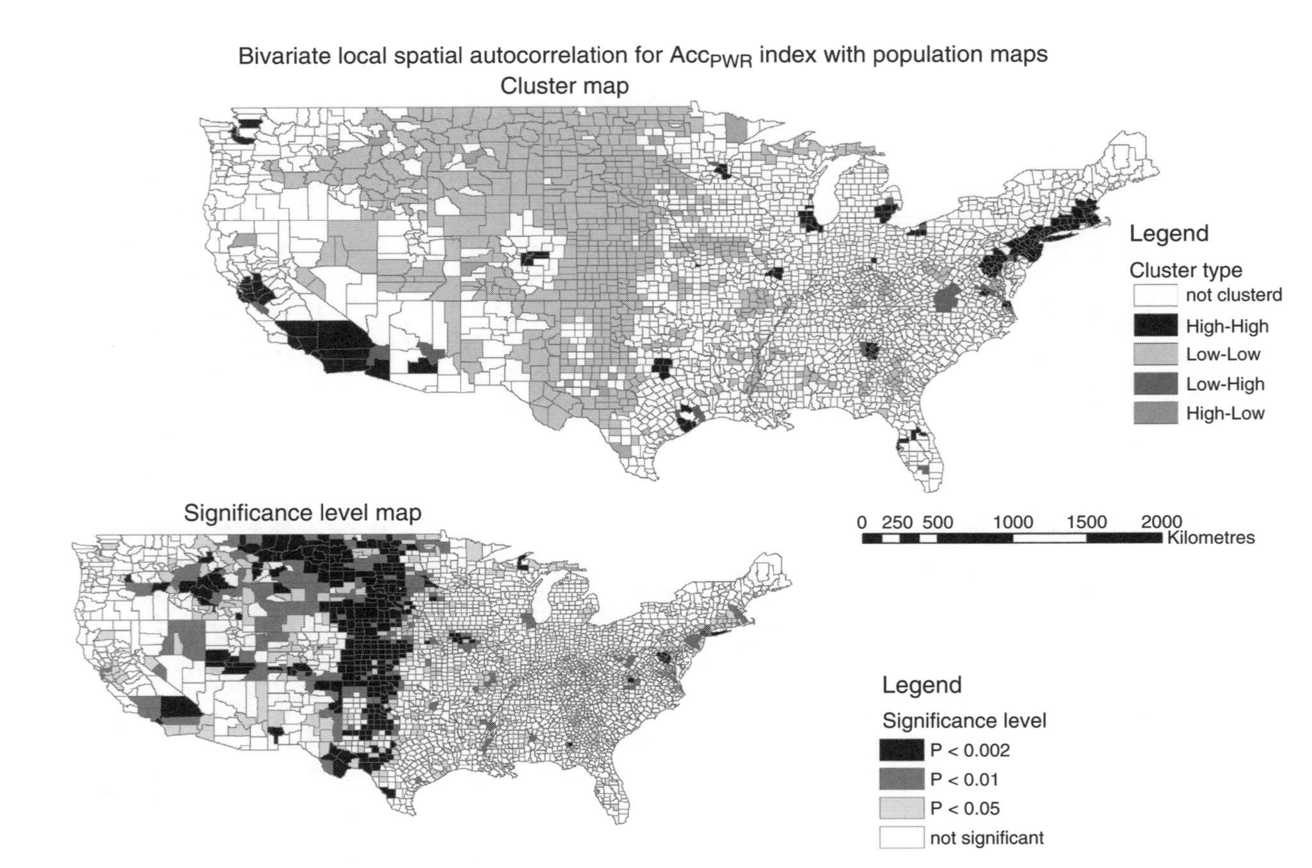

Figure 6.6 LISA bivariate cluster map for US counties for residential population versus accessibility (top) and significance level map (bottom)

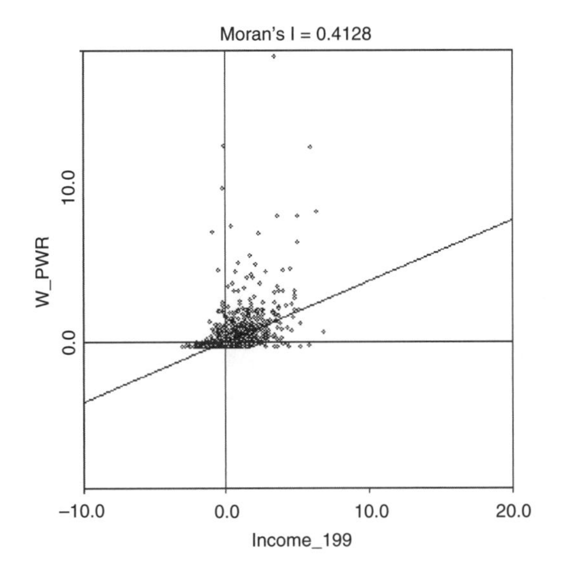

Figure 6.7 *Spatial bivariate global autocorrelation analysis of accessibility versus income per capita*

6.6 CONCLUSION: DISCUSSION AND FUTURE LINES OF RESEARCH

In this section, we give our interpretation of the results presented in this chapter and discuss issues which call for further research. From a general perspective, the geography, infrastructure and logistical features of an area influence accessibility. The spatial autocorrelation analyses implemented in this study have brought forth some interesting and not always expected findings.

Univariate autocorrelation analysis for county accessibility and bivariate autocorrelation analysis of accessibility and population display very similar patterns. The most accessible counties are counties that have a high population density. A closer look at Figures 6.5 and 6.6 reveals that large clusters composed of highly accessible counties are located on the north-east coast (70 counties) and in the west on the Californian coast (20 counties), while another eight small clusters (composed of six counties on average) are spread around the rest of the US. It is not surprising that all these clusters have developed around one or more large cities: Boston, New York, Philadelphia, Baltimore and Washington, DC (in the large cluster on the north-east coast), San Francisco and Los Angeles (on the west coast), Detroit, Chicago, St Louis, Atlanta, Virginia Beach, Denver,

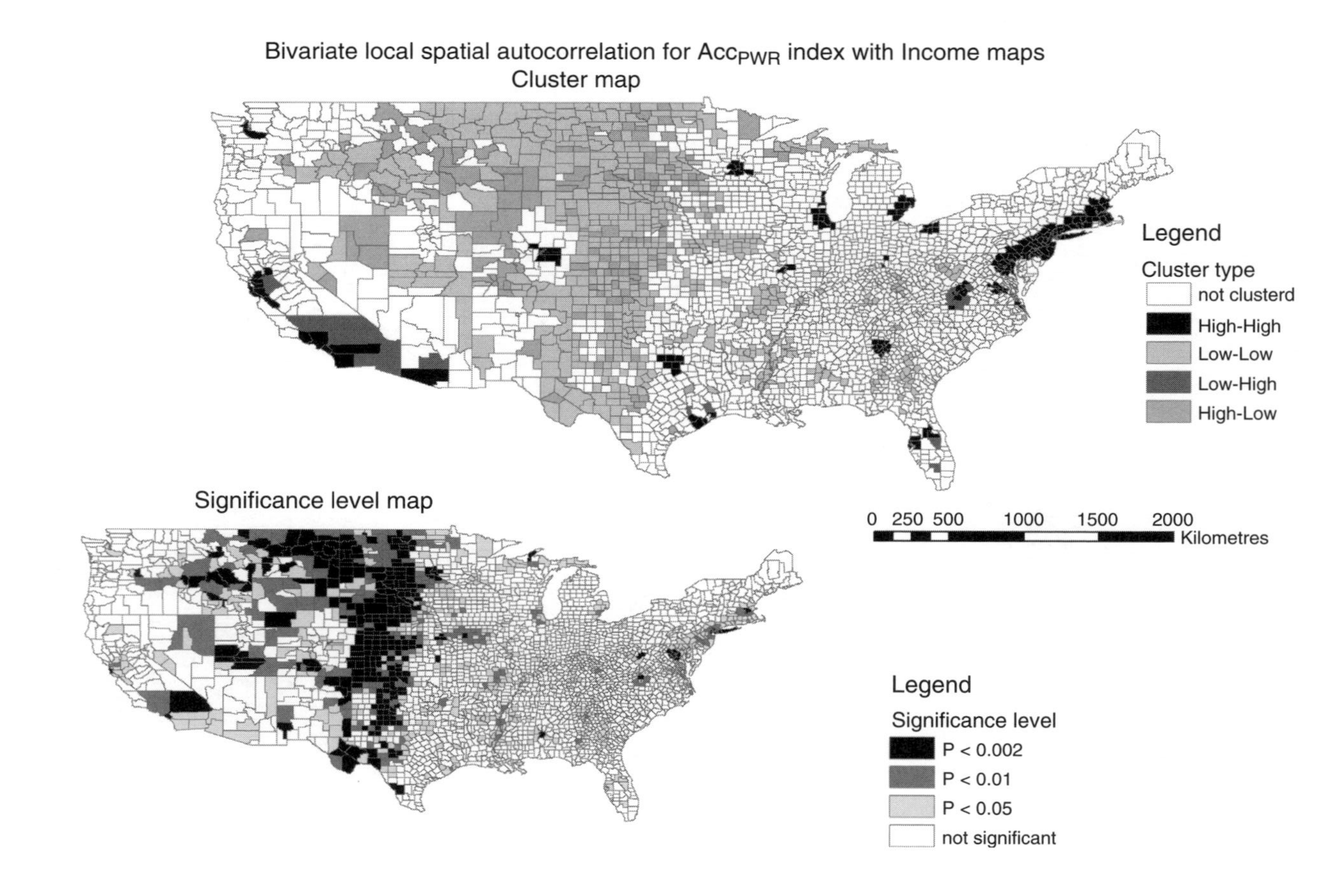

Figure 6.8 LISA bivariate cluster map for US counties for average income per capita versus accessibility (top) and significance level map (bottom)

Houston and Dallas (in the rest of the US). When we look at the rest of the nation, another important picture emerges: a large portion of the central states (North Dakota, South Dakota, Nebraska, west Kansas, Oklahoma, New Mexico, Wyoming, Montana, Idaho, Utah and north Texas) form a homogeneous cluster composed of low accessibility counties. These are deep rural areas, characterized by low population density, low economic performance and higher poverty rates (Porter et al., 2004). In this sense, this analysis confirms that the accessibility of a location is not only due to the characteristics of a given territorial unit, but is also closely connected to the level of development (or underdevelopment) of neighbouring areas. As we have commented in previous sections, the average commuting trip in the US is within a distance radius of 25 minutes. This implies that the majority of commuting trips take place within the county of residence or to and from nearby counties. This is supported by the evidence that clusters which are highly accessible (shown in black in Figure 6.4) are composed of counties with a small to medium area. In such counties it is easier to have an inter-county flow of commuters, because users have to travel less to cross the border of their county of residence. The same picture also emerges in the Los Angeles cluster. In this case, the population is concentrated around the Los Angeles metropolitan area, which includes the counties of Los Angeles, San Bernardino, Orange County and Riverside. Although these counties have a large total area, only the sections of the counties located within the Los Angeles metropolitan area are highly accessible. We can conclude that in this case the measure of accessibility does not really reflect the total accessibility of the four counties but only of the Los Angeles metropolitan area (a dense homogeneous urban part of the four counties).

Regarding the spatial bivariate correlation between accessibility and average income, we have found a high similarity with the population map for the 'high–high' and 'low–low' cluster types. It is interesting to note the case of the 'high–low'cluster type. These counties have a low accessibility and are surrounded by counties which have high average income. This phenomenon occurs almost exclusively in the central rural part of the US (Figure 6.8) and can be explained as follows. In these counties, the economy is grounded on relatively richer inhabitants who are marginally affected by commuting.

The arguments presented in this chapter have focused on a number of issues and left some of them at least partially unanswered. Regarding the question of accessibility for counties with fairly large areas, we recognize the need for further in-depth analysis. This is a reflection of the well known 'modifiable areal unit problem' described by Openshaw and Taylor (1979, 1981). In our case a geographical bias is unavoidably introduced, because spatial data has not been corrected for the case of small counties with high commuting flows (especially in the north-east).

Disparity in spatial development is yet another issue we have identified. There is a diverging pattern in the distribution of accessibility. Relatively small regions benefit from high levels of accessibility, partially explained by high levels of population and income per capita; while large regions lag behind, showing reduced levels of accessibility and socio-economic variables. This scenario leads to a number of issues, such as the role that rurality and accessibility/remoteness will play in future US socio-economic development, and the need for policies related to infrastructure quality in spatial and transport planning. These aspects deserve further research in the future.

A third issue has to do with methodology. In this chapter, we have applied bivariate spatial autocorrelation analyses in order to explain the spatial distribution of accessibility. This approach may be compared to similar techniques, such as geographically weighted regression, which are able to assess the influence of certain spatial phenomena on the distribution of accessibility in different counties throughout the US. In this framework, beyond population and income, the set of explanatory variables may include further descriptors whose choice depends on the specific issues and policies under scrutiny.

NOTES

1. NUTS is a nomenclature introduced by Eurostat for referencing the administrative subdivisions of countries for statistical purposes.
2. Source: US Census Bureau.
3. Source: Eurostat.

REFERENCES

Anselin, L. (1995), 'Local indicators of spatial autocorrelation – LISA', *Geographical Analysis*, **27** (2), 93–115.

Anselin, L. (2003), 'An introduction to spatial autocorrelation analysis with GeoDa', available at http://www.utdallas.edu/~briggs/poec6382/geoda_spauto.pdf (accessed 26 July 2013).

Baradaran, S. and F. Ramjerdi (2001), 'Performance of accessibility measures in Europe', *Journal of Transportation and Statistics*, **4** (2–3), 31–48.

Caschili, S. and A. De Montis (2013), 'Accessibility and complex network analysis of the US commuting system', *Cities*, **30**, 4–17.

Cliff, A.D. and J.K. Ord (1969), 'The problem of spatial autocorrelation', in A.J. Scott (ed.), *London Papers in Regional Science 1, Studies in Regional Science*, London: Pion, pp. 25–55.

Curry, L. (1972), 'A spatial analysis of gravity flows', *Regional Studies*, **6** (2), 137–147.
De Montis, A., M. Barthélemy, A.Chessa and A. Vespignani (2007), 'The structure of interurban traffic: a weighted network analysis', *Environment and Planning B: Planning and Design*, **34** (5), 905–924.
De Montis, A., S. Caschili and A. Chessa (2011), 'Time evolution of complex networks: commuting systems in insular Italy', *Journal of Geographical Systems*, **13** (1), 49–65.
De Montis, A. and A. Reggiani (2012), 'Special section on accessibility and socio-economic activities: methodological and empirical aspects', *Journal of Transport Geography*, **25**, 95–97.
De Montis, A. and A. Reggiani (2013), 'Special section on analysis and planning of urban settlements: the role of accessibility', *Cities*, **30**, 1–3.
Fingleton, B. (2006), 'A cross-sectional analysis of residential property prices: the effects of income, commuting, schooling, the housing stock and spatial interaction in the English regions', *Papers in Regional Science*, **85** (3), 339–361.
Fotheringham, A.S. and M.E. O'Kelly (1989), *Spatial Interaction Models: Formulations and Applications*, Dordrecht: Kluwer Academic.
Getis, A. (2007), 'Reflections on spatial autocorrelation', *Regional Science and Urban Economics*, **37** (4), 491–496.
Geurs, K.T. and B. van Wee (2004), 'Accessibility evaluation of land-use and transport strategies: review and research directions', *Journal of Transport Geography*, **12** (2), 127–140.
Gillen, K. (2001), 'Anisotropic autocorrelation in house prices', *Journal of Real Estate Finance and Economics*, **23** (1), 5–30.
Griffith, D.A. (2007), 'Spatial structure and spatial interaction: 25 years later', *Review of Regional Studies*, **37** (1), 28–39.
Griffith, D.A. (2009), 'Spatial autocorrelation', *International Encyclopedia of Human Geography*, Amsterdam, the Netherlands and London, UK: Elsevier Science, pp. 1–10.
Handy, S.L. and D.A. Niemeier (1997), 'Measuring accessibility: an exploration of issues and alternatives', *Environment and Planning A*, **29** (7), 1175–1194.
Hansen, W.G. (1959), 'How accessibility shapes land use', *Journal of the American Institute of Planners*, **25** (2), 73–76.
Horner, M.W. (2008), 'Spatial dimensions of urban commuting: a review of major issues and their implications for future geographic research', *Professional Geographer*, **56** (2), 160–173.
Jones, S.R. (1981), 'Accessibility measures: a literature review', TRRL Report 967, Crowthorne: Transport and Road Research Laboratory.
Kawabata, M. and Q. Shen (2007), 'Commuting inequality between cars and public transit: the case of the San Francisco Bay Area, 1990–2000', *Urban Studies*, **44** (9), 1759–1780.
Lenormand, M., S. Huet, F. Gargiulo and G. Deffuant (2012), 'A universal model of commuting networks', *PLOS ONE*, **7** (10), e45985.
Lucy, W. and D. Phillips (1997), 'The post-suburban era comes to Richmond: city decline, suburban transition and exurban growth', *Landscape and Urban Planning*, **36** (4), 259–275.
Martín, J.C. and A. Reggiani (2007), 'Recent methodological developments to measure spatial interaction: synthetic accessibility indices applied to high-speed train investments', *Transport Reviews*, **27** (5), 551–571.

McMillen, D.P. (2004), 'Employment densities, spatial autocorrelation, and subcenters in large metropolitan areas', *Journal of Regional Science*, **44** (2), 225–243.
Molho, I. (1995), 'Spatial autocorrelation in British unemployment', *Journal of Regional Science*, **35** (4), 641–658.
Moran, P.A.P. (1950), 'Notes on continuous stochastic phenomena', *Biometrika*, **37** (1), 17–23.
Openshaw, S. and P. Taylor (1979), 'A million or so correlation coefficients: three experiments on the modifiable area unit problem', in N. Wrigley (ed.), *Statistical Applications in the Spatial Sciences*, London: Pion, pp. 127–144.
Openshaw, S. and P. Taylor (1981), 'The modifiable areal unit problem', in N. Wrigley and R.J. Bennett (eds), *Quantitative Geography: A British View*, London: Routledge, pp. 60–69.
Páez, A. and D.M. Scott (2004), 'Spatial statistics for urban analysis: a review of techniques with examples', *GeoJournal*, **61**, 53–67.
Patacchini, E. and Y. Zenou (2007), 'Spatial dependence in local unemployment rates', *Journal of Economic Geography*, **7** (2), 169–191.
Porter, M.E., C.H.M. Ketels, K. Miller and R.T. Bryden (2004), *Competitiveness in Rural US Regions: Learning and Research Agenda*, Cambridge, MA: Institute for Strategy and Competitiveness, Harvard Business School.
Reggiani, A. (2008), 'Networks and accessibility structures: German commuting patterns', in U. Becker, J. Böhmer and R. Gerike (eds), *How to Define and Measure Access and Need Satisfaction in Transport*, Dresden: Dresden Institute for Transportation and Environment (DIVU), University of Dresden, pp. 95–108.
Reggiani, A. (2011), 'Accessibility and resilience in complex networks', paper presented at West Virginia University, USA, 28 April, available at rri.wvu.edu/wp-content/uploads/2011/04/Paper_for_Reggiani_Accessibility_and_Resilience.pdf.
Tobler, W. (1970), 'A computer movie simulating urban growth in the Detroit region', *Economic Geography*, **46** (2), 234–240.
Van Ham, M., P. Hooimeijer and C. Mulder (2001), 'Urban form and job access: disparate realities in the Randstad', *Tijdscrift voor Economische en Sociale Geographie*, **92** (2), 231–246.
Vandenbulcke, G., C. Dujardin, I. Thomas, B. de Geus, B. Degraeuwe, R. Meeusen and L.I. Panis (2011), 'Cycle commuting in Belgium: spatial determinants and "re-cycling" strategies', *Transportation Research Part A*, **45** (2), 118–137.
Wang, F. (2001), 'Explaining intraurban variations of commuting by job proximity and workers' characteristics', *Environment and Planning B: Planning and Design*, **28** (2), 169–182.
Weibull, J. (1976), 'An axiomatic approach to the measurement of accessibility', *Regional Science and Urban Economics*, **6** (4), 357–379.
Wilson, A.G. (1971), 'A family of spatial interaction models and associated developments', *Environment and Planning A*, **3** (1), 1–32.
Wu, B.M. and J.P. Hine (2003), 'A PTAL approach to measuring changes in bus service accessibility', *Transport Policy*, **10** (4), 307–320.

7. Border effect and market potential: the case of the European Union

María Henar Salas-Olmedo, Ana Condeço-Melhorado and Javier Gutiérrez*

7.1 INTRODUCTION

Access to international markets is influenced by the border effect. Despite all efforts to reduce barriers to trade and increase international commercial flows, several studies confirm that borders still matter from the commercial trade viewpoint, even within international institutions that form a single market, such as the European Union. Whilst in recent years a growing literature has analysed the decrease in volume of trade due to national borders, most accessibility studies at the supranational level ignore border effects, thus obtaining unrealistic results. One exception is the article by Gutiérrez et al. (2011), who looked at international accessibility between European regions and lowered the weight of international destinations by a factor of ten, based on the results of previous studies on the border effect in the European Union.

Previous research on border effects reports an excess of national trade when compared to what would have been traded with an otherwise identical foreign country. This effect, also known as home bias, has intrigued several researchers since McCallum's (1995) seminal study was published.

McCallum applied gravity-type equations to test the relevance of the US–Canadian border for regional trade. He assumed that trade within regions of the same country is a function of each region's gross domestic product (GDP) and the distance between them. He added a dummy variable equal to 1 for intra-national trade and 0 for trade between one Canadian province and one US state. This variable represented the border effect between the two countries and he found that, on average, trade between Canadian provinces is about 22 times higher than their rate of trade with other US states. This is surprising, particularly considering that the two countries share a common language, have similar cultural and institutional characteristics, and have systematically reduced barriers to trade over the years.

Subsequent studies have attempted to validate the existence of border effects, obtaining different results. From the imports perspective, Wei (1996) provided a worldwide analysis of the home bias as well as results for sets of countries involved in regional trade unions. In addition to McCallum's variables, he tested the significance of the location of the country relative to all others (that is, remoteness) and two dummy variables (adjacency and common language). His results evidenced a far lower border effect than McCallum's figures. For Organisation for Economic Co-operation and Development (OECD) countries the simple gravity model indicated a home bias of 9.7, whereas the inclusion of remoteness, adjacency and language variables reduced this to 2.6. He also observed a declining trend in the border effect between 1982 and 1994. Wei computed the model over a sample of eight members of the then European Community (EC), evidencing that domestic trade in European countries was only 1.7 times higher than trade with other EC members.

Clark and van Wincoop (2001) compared intra-national and international border effects and synchronization with business cycles in the US and the European Union (EU), obtaining results in line with those of Wei (1996). They reported a very slight, not even statistically significant, reduction of the border effect in European countries between 1964 and 1997 (from 1.32 in 1964–1980, to 1.27 in 1981–1997).

Focusing on the case of Europe, Nitsch (2000) concluded that in 1982 European countries traded between seven and ten times more within their own borders than with other EU members. He obtained these figures after controlling for country size, distance and language, obtaining different values for alternative sets of countries. Head and Mayer (2002) analysed the border effect after implementation of the Single Market Programme using a variation of the distance measurement, which they defined as the effective distance, which lowered the estimation of the border effect to 4.2. Chen (2004) limited her analysis to a set of seven EU countries and to the manufacturing sector. She compared several distance metrics (all of them based on great circle distances) and established that in the manufacturing sector an EU country trades six times more with itself than with another EU country.

One of the causes underlying different results in the previous literature seems to be differences in distance metrics. Although the influence of distance metrics in the final outcome of the border effect has already been pointed out in several studies, most authors eventually use Euclidean or great circle distances, thus failing to provide a realistic measure of the distance variable. This chapter contributes to the border effect literature by analysing the influence of different distance metrics on border effect estimations. We follow the afore-mentioned branch of the literature that measures the border effect using a gravity model and focus on the European Union. Despite being criticized for its weak theoretical foundations (Anderson and van Wincoop, 2001), the gravity model forms the basis of a strong empirical background for dealing with the explanation of the border effect in international trade. We compare simple distance measurements (that is, Euclidean) with more realistic measurements such as network distance, travel time and generalized transport cost (GTC). Border effects are calculated for the EU as a whole and for a set of individual countries in order to draw conclusions about their spatial variation across countries.

In this chapter, our emphasis is on providing an analysis of the border effect, thus enabling its integration into further accessibility estimations on an international scale. This study contributes to the accessibility literature by testing the effect of including the border effect in an accessibility indicator. The existence of border effects would reduce the weight of international relationships in accessibility analysis. According to previous literature, this effect would eventually differ across countries as countries evidence different border effects on international trade.

The rest of the chapter is structured as follows: section 7.2 reviews how distance is measured in previous border effect studies; section 7.3 describes the data and methods; section 7.4 presents the main results and the discussion of our border effect estimations followed by an example of its impact on accessibility analysis; and section 7.5 contains our conclusions.

7.2 MEASUREMENT OF INTERNATIONAL AND NATIONAL DISTANCES IN BORDER EFFECT GRAVITY MODEL ESTIMATIONS

Distance increases transport costs, thus imposing a barrier to trade. Consequently, gravity models negatively relate trade with increased distance. Some studies found that the border effect is sensitive to how national and international distances are measured. While international distances are usually measured as great circle or Euclidean distances between centres (usually capitals, large cities or simply centroids), national distances are generally estimated as distances to centres of neighbouring countries or as a function of the area of each country (Head and Mayer, 2002).

One example is the work by Wei (1996), who conceptualized international trading distances as the great circle distance between the main city of each country, and internal distance as one-quarter of the distance between the capital city and the closest capital city of a neighbouring country. Wei's measurement of distance relied on several assumptions that are far from reality. First, he assumed that the economic activity of a country is concentrated in its capital, which is particularly unrealistic in large countries with several important cities. Second, Wei's measure of internal distance assumed that both capital cities are equally distant from the border, which is rarely the case for many countries, as Nitsch (2000) pointed out. Finally, he did not take into account the size of the country; thus, two neighbouring countries of different size might end up with the same internal distance. Nitsch criticized the use of straight point-to-point measurements within countries such those used by Wei, and proposed instead estimating internal distances as a function of the size of each country.

Several studies concluded that the method used for measuring both interregional and intra-regional distances affects the final magnitude of border effects. Wei highlighted the influence of intra-national distance measurement estimations when checking the robustness of his model. He found that increasing (decreasing) intra-national distance by 25 per cent greatly affected the magnitude of the border effect. This fact was also acknowledged by Hillberry and Hummels (2003) who concluded that overstating distances, and particularly intra-regional distances, increases the magnitude of border effects. Head and Mayer (2002) recalled the importance of applying the same methodology to inter- and intra-national distances in order to avoid an overestimation of the latter.

Despite all the studies that analysed the impact of using different distance metrics on the estimation of the border effect (see Table 7.1), previous research mostly measured distances in a very simple way. Some studies used capital cities as representative of countries, from which distances to other countries (also represented as their capital city) are measured. These studies assumed that all economic activity is located in the capital city, thus leading to inaccurate estimations of both inter- and intra-national distances. In addition, the use of Euclidean and great circle distances is an oversimplification because trade is shipped using the transport infrastructure available. Nowadays network distances and travel times can be calculated using network analysis tools included in geographical information systems (GIS). Furthermore, generalized transport costs can be measured with specific tools for transport modelling. According to the *Statistical Pocketbook* (European Union, 2012), in the EU27 the most common mode in terms of freight transport (tonne–kilometres) is road transport (42.1 per cent), followed by maritime shipping (37.5 per cent) and rail (12.6), whereas only a small proportion of freight is transported by other modes. This variation across modes can only be captured by using more complex measurements of distance such as generalized transport costs.

Table 7.1 Review of internal distance measurements

Author (s)	Type of measurement	Spatial unit (coverage)	Implementation	Source
Wei (1996)	Great circle distance	Country (worldwide)	¼ of the distance between one country's main city and that of the closest country	Geometric
Wolf (1997, 2000)	Network based	State (US)	1. the distance between the largest and the second largest city within the state; 2. half the distance to the largest city in all adjacent states	Road atlas
Hillberry and Hummels (2002)	Network based	State (US)	Reported shipped distance	US Commodity Flow Survey
Nitsch (2000)	Function of area	Country (EU)	$1/\sqrt{\pi}$ (= 0.56) times the square root of the area of the country	Geometric
Helliwell and Verdier (2001), Helliwell (2002)	Function of area	Province (Canada)	$0.52\sqrt{\text{area of the province}}$	Geometric
Head and Mayer (2002)	Great circle distance	State (US) and country (EU)	$d_{ij} = \left(\sum_{k \in i} (y_k/y_i) \sum_{l \in j} (y_l/y_j) d_{kl}^{\theta} \right)$	GDP and geometric distance between aggregation units (k,l) lower than the target units (i,j)
Chen (2004)	Great circle distance	Country (EU)	$D_{ij} = \frac{\sum (D_{m_i m_j} \times S_{m_i} \times S_{m_j})}{\sum (S_{m_i} \times S_{m_j})}$ $S_m = \left(\frac{GDP_m}{GDP} \right)$	GDP and geometric distance between aggregation units (m) lower than the target units (i,j)

One of our main contributions made by this study has been to improve the measurement of distance specifications in gravity models, comparing network distance, travel time and GTC. In addition, we also compute Euclidean distances in order to compare our results with previous studies.

7.3 MEASURING THE BORDER EFFECT

7.3.1 Data

In this chapter, as in other European studies on border effects (Chen, 2004; Nitsch, 2000), Eurostat was chosen as the main data source in order to obtain the most homogeneous and reliable data for the EU27 countries.[1] Eurostat's Comext and Structure Business Statistics (SBS) are the recommended datasets for the construction of a matrix containing the value in euros of commercial flows between EU27 countries inclusive of the internal trade in each country. From Comext we obtained country-to-country export flow data as well as the total exports of each EU27 country to the rest of the world. The SBS provided information on the value in euros of the national production. We analysed manufactured goods in 2009, thus taking advantage of the most recent and complete production datasets available. Total exports and national production figures allowed us to estimate internal trade by calculating the national production minus exports to all countries (as Chen 2004) (suggested following previous studies). This data was used to complete the diagonal of the matrix. Contrary to Chen and others, we did not analyse subcategories of manufactured commodities. This allowed us to increase the spatial coverage of our analysis.

The combination of different datasets caused some limitations that affected our results. National production was collected by the corresponding national statistics institutes through surveys to firms. We therefore compared two slightly different datasets: on the one hand, international trade of manufactured products, and on the other hand, national production of firms classified in the manufacturing sector. In addition, two methodologies are applied to the former since international trade is collected from customs records through the Intrastat methodology for intra-EU27 arrivals and dispatches, and through the ExtraStat methodology for imports and exports to the rest of the world (details on these methodologies are available from the European Commission (2006).

Whilst the afore-mentioned data mismatches did not affect the core of the analysis, the so-called 'Rotterdam effect'[2] did reveal an important issue concerning the analysis of European trade statistics (European Commission, 2006, p. 9). We will now explain its origin and implications for this study.

Ideally, domestic trade equals national production minus the part of national production that is exported (that is, we need to remove the exports of imported products), plus the imported products that remain in the country. Unfortunately, current databases do not indicate whether exports (or imports) come from (go to) national production or from (to) third countries. This poses a problem of double-counting intermediate consumption when totalling imports and national production. Therefore, some authors suggest that domestic trade equals national production minus exports (Wei, 1996; Nitsch, 2000; Head and Mayer, 2000; Chen, 2004). However, some countries import high volumes of commodities from extra-EU countries that are then exported to EU countries by local firms favoured by the current free trade framework, which in turn questions the assumption that domestic trade equals national production minus exports. Therefore, we need to agree on a trade-off between the real concept of domestic trade, data availability and the spatial coverage of the analysis.

In this study, we tried to identify and exclude from the analysis those countries particularly affected by the 'Rotterdam effect' by selecting those in which exports include a high proportion of imported products. Table 7.2 shows that imports and exports of Cyprus and Luxembourg, respectively, exceed national production, thus they were automatically excluded from our study. We also excluded from the study those countries in which the percentage of imports or exports over the national production is larger than the average plus one standard deviation of the remaining 24 countries. In these countries (Belgium, Estonia, Ireland, Lithuania, Latvia and the Netherlands), the proportion of domestic trade is below 30 per cent, whereas the average of the remaining 18 countries doubles this percentage (60 per cent).

In view of the data shown in Table 7.2, the Rotterdam effect is particularly strong in the Netherlands, Belgium and Luxembourg, as well as, to a lesser extent, Ireland, Estonia, Latvia and Lithuania. For the rest of the countries, and in the absence of direct data, the assumption that domestic trade equals national production minus exports provides an acceptable estimation.

We used the TRANS-TOOLS road and ferry link network to compute network distances (km and travel time) between all NUTS 2 centroids[3] using the network analysis tools from a commercial GIS software (ArcGIS 10). Alternatively, we took the generalized transport cost (GTC) in euros per tonne of all modes of transport from TRANS-TOOLS.[4]

Table 7.2 Estimation of domestic manufacturing trade, 2009

Country	Exports (m euros)	Imports (m euros)	National production (m euros)	Domestic trade (m euros)*	Domestic trade (%)	Imports (%)	Exports (%)
Austria (AT)	59 551	62 097	133 731	74 180	55.5	46.4	44.5
Belgium (BE)	183 965	156 566	198 556	14 592	7.3	78.9	92.7
Bulgaria (BG)	8 558	9 379	19 025	10 467	55.0	49.3	45.0
Cyprus (CY)	459	4 142	3 322	2 863	86.2	124.7	13.8
Czech Republic (CZ)	46 253	49 631	107 959	61 706	57.2	46.0	42.8
Germany (DE)	439 168	372 755	1 378 217	939 049	68.1	27.0	31.9
Denmark (DK)	42 041	38 091	76 292	34 252	44.9	49.9	55.1
Estonia (EE)	4 514	5 438	5 702	1 188	20.8	95.4	79.2
Spain (ES)	94 093	120 828	394 143	300 050	76.1	30.7	23.9
Finland (FI)	31 384	24 370	93 512	62 128	66.4	26.1	33.6
France (FR)	200 307	233 565	678 510	478 203	70.5	34.4	29.5
Great Britain (GB)	146 256	230 359	462 846	316 590	68.4	49.8	31.6
Greece (GR)	10 764	28 586	50 150	39 386	78.5	57.0	21.5
Hungary (HU)	37 636	34 687	65 408	27 772	42.5	53.0	57.5
Ireland (IE)	68 759	28 719	93 747	24 988	26.7	30.6	73.3

Italy (IT)	175 435	168 205	742 725	567 291	76.4	22.6	23.6
Lithuania (LT)	8 305	6 793	11 425	3 121	27.3	59.5	72.7
Luxembourg (LU)	12 488	12 479	7 837	−4 651	−59.3	159.2	159.3
Latvia (LV)	3 619	4 721	4 777	1 159	24.3	98.8	75.7
Malta (MT)	1 675	2 315	n.a.	–	–	–	–
Netherlands (NL)	228 956	188 221	229 346	389	0.2	82.1	99.8
Poland (PL)	60 394	68 032	178 536	118 141	66.2	38.1	33.8
Portugal (PT)	23 043	31 118	65 803	42 760	65.0	47.3	35.0
Romania (RO)	16 934	26 267	45 142	28 208	62.5	58.2	37.5
Sweden (SE)	61 796	52 718	138 352	76 556	55.3	38.1	44.7
Slovenia (SI)	11 369	12 224	18 550	7181	38.7	65.9	61.3
Slovakia (SK)	25 586	25 100	40 328	14 742	36.6	62.2	63.4

Note: * Domestic trade = National production – Exports.

Source: Eurostat.

7.3.2 Methods

We considered the gravity model chosen by Chen in her study of the border effect of manufactured commodities between seven European countries in order to obtain a parameter of the border effect that could be further integrated in accessibility indicators. There are two main differences with respect to the original work of McCallum (1995) that make this model more suitable for our study. First, Chen substituted the GDP of the origin country with the value of the national production in manufactures. Similar to our case, this is in accordance with the specific sector of the economy that she was analysing.

The second change was the introduction of a new dummy variable, that is, country adjacency. The dummy *home* already accounts for domestic trade, therefore in this new variable domestic trade is computed as non-adjacent. Adjacency is commonly integrated in the gravity model when the research includes a large number of countries, as is the case of European or worldwide research (Chen, 2004; Nitsch, 2000; Wei, 1996). This dummy variable might be interpreted as a proxy for cultural and historical relationships between neighbouring countries. In addition, from a freight cost perspective neighbouring countries are usually better connected, particularly via train and highway, which generates a much closer relationship between neighbouring countries. Following Chen, and contrary to some other authors, we did not include a third dummy variable to account for common language, thus our results are easily comparable to hers.

The following equation expresses Chen's specification of the gravity model (Chen, 2004, p. 95):

$$\ln X_{ij} = \beta_0 + \beta_1\, home + \beta_2 \ln Y_i + \beta_3 \ln Y_j + \beta_4 adj_{ij} + \beta_5 \ln D_{ij} + \varepsilon_{ij} \quad (7.1)$$

where X_{ij} is trade between exporting country i and importing country j expressed in euros, *home* is a dummy equal to 1 for domestic trade and 0 otherwise, Y_i is the value in euros of the manufacturing production of the exporter country i, Y_j is the GDP of the importing country j, adj_{ij} is a dummy equal to 1 when country i and j share a common border and 0 otherwise, and D_{ij} is the distance between countries i and j. Finally, our analysis did not require the use of the Tobit procedure due to the completeness of the flow matrix (that is complete lack of zero values).

Following Chen (2004, p. 117), we applied a common methodology to compute intra-national and international distances weighted by origin and destination GDP. First, we calculated the share (S_m) of the GDP of region m in its country GDP according to:

$$S_m = \left(\frac{GDP_m}{GDP}\right) \tag{7.2}$$

Then, we calculated the distance between the exporter country i and the importer country j (D_{ij}) as a weighted average of the distance between all NUTS 2/3 regions (as indicated in section 7.3.1) in origin country i and all NUTS 2/3 regions in destination country j:

$$D_{ij} = \frac{\sum (D_{m_i m_j} \times S_{m_i} \times S_{m_j})}{\sum (S_{m_i} \times S_{m_j})} \tag{7.3}$$

where $D_{m_i m_j}$ is the value of the distance between each region m of country i and each region m of country j.

The value of D_{ij} was calculated in terms of Euclidean distance, network distance, travel time and GTC in order to set a framework for the analysis of the role of distance measurement in the gravity model. Equation (7.3) also applies for intra-national distances (when country i and j are the same).

The dummy *home* is our focus variable and its antilogarithm evidences the magnitude of the home bias, that is, the excess of national trade when compared to what would have been traded with an otherwise identical foreign country. Wei (1996, p. 3) defined the border effect as 'an all-inclusive summary of "barriers" to trade'. In our analysis, border effects account for all international barriers to trade that might exist in the EU and anything else that is not included in the list of control variables that may contribute to intra-national, but not international, trade. The size of the border effect has an impact on accessibility levels since it implies an additional reduction of trade between foreign regions.

7.4 BORDER EFFECT AND MARKET POTENTIAL

7.4.1 The Border Effect in the EU: Discussion of Main Results

The analysis was performed both at the global (Europe) and national level. First, linear regressions (ordinary least squares, OLS) were computed using different distance measurements in order to show the impact of using different distance specifications on border effect estimations. Additionally, three different sets of countries were used in order to estimate the impact of the Rotterdam effect on the home bias at a European level. Then, individual analysis per country was completed using different distance measures.

Unlike previous studies, a complete dataset of the required variables is now available for all EU27 countries, with the only exception of national production for Malta (see Table 7.2). Table 7.3 shows how the selection of the set of countries affects the global border effect. In the first column, only Cyprus, Luxembourg and Malta (for which internal distances cannot be computed, as explained in section 7.3.1) are excluded from the sample. In the second and third columns, countries largely affected by the Rotterdam effect were subsequently excluded from the analysis.

High adjusted R^2 values were obtained in all the regressions, with values near or over 0.9, all of them being significant at the 0.01 level. All predictors had the expected coefficient signs and lack multicollinearity (VIF < 7.5, not shown in the table), which were also significant at the 0.01 level (with the exception of adjacency in some specifications). A positive coefficient of the dummy variable *home* suggests a preference for trading within the country rather than with other countries. The antilog of this coefficient (last line in each specification) shows the size of the border effect, which increased with the complexity of the distance measurement method as well as with the exclusion of countries with identified Rotterdam effect.

Focusing on the set of countries less affected by the Rotterdam effect (third column in Table 7.3), results evidenced an increment of the border effect from 6.4 with Euclidean distance to 9.3 with the network distance and 15 using travel time or GTC. This result is in line with the findings of Helliwell (2002) and Chen (2004), which show that the border effect increases with the complexity of the distance measurement method. Unlike these two researchers, we introduced more realistic distance metrics such as network distance, travel time and GTC. Each step towards a more realistic approach to transport cost provided a considerable increase in the border effect, thus suggesting that previous studies using Euclidean and great circle distances underestimated it.

The percentage of exports over the national production for countries clearly affected by the Rotterdam effect are overestimated, which leads to an underestimation of the border effect. According to GTC figures, European countries seem to trade within themselves almost six times more than with the other countries in the dataset when removing countries the most affected by the Rotterdam effect (that is Belgium and the Netherlands). This bias is higher (15 times) when removing all the countries that showed evidence of the Rotterdam effect. Therefore, the border effect seems to be highly sensitive to both the distance metric used and the set of countries considered.

Despite some differences between our model and other studies (for example, different set of countries, sectors analysed, distance conceptualizations, or variables included in the gravity model), our results for

Table 7.3 Global border effect of the manufacturing sector in the European Union, 2009

	EU27 (excl. CY LU MT)	EU27 (excl. BE CY LU MT NL)	EU27 (excl. BE CY EE IE LT LU LV MT NL)
Euclidean distance			
home	0.795*	1.061*	1.869*
Ln Distance ij	−1.520*	−1.620*	−1.372*
Ln Production i	0.860*	0.861*	0.860*
Ln GDP j	0.792*	0.804*	0.797*
Adj ij	0.214**	0.216**	0.360*
Observation	24 × 24	22 × 22	18 × 18
S.E.R.	0.726	0.635	0.522
R^2	0.893	0.921	0.931
Adjusted R^2	0.892	0.920	0.930
Border effect [exp. home]	2.215	2.889	6.485
Network distance			
home	1.238*	1.486*	2.230*
Ln Distance ij	−1.567*	−1.708*	−1.403*
Ln Production i	0.855*	0.860*	0.843*
Ln GDP j	0.790*	0.805*	0.785*
Adj ij	0.199	0.164	0.351**
Observation	24 × 24	22 × 22	18 × 18
S.E.R.	0.717	0.610	0.515
R^2	0.896	0.927	0.933
Adjusted R^2	0.895	0.926	0.932
Border effect [exp. home]	3.450	4.419	9.301
Travel time			
home	1.746*	1.993*	2.710*
Ln Distance ij	−1.325*	−1.456*	−1.172*
Ln Production i	0.818*	0.817*	0.818*
Ln GDP j	0.753*	0.762*	0.761*
Adj ij	0.419**	0.388**	0.566*
Observation	24 × 24	22 × 22	18 × 18
S.E.R.	0.796	0.720	0.578
R^2	0.872	0.898	0.916
Adjusted R^2	0.871	0.897	0.915
Border effect [exp. home]	5.729	7.340	15.033
GTC			
home	1.484*	1.790*	2.751*
Ln Distance ij	−1.650*	−1.786*	−1.309*
Ln Production i	0.879*	0.882*	0.851*
Ln GDP j	0.814*	0.831*	0.834*
Adj ij	0.310**	0.291**	0.581*

Table 7.3 (continued)

	EU27 (excl. CY LU MT)	EU27 (excl. BE CY LU MT NL)	EU27 (excl. BE CY EE IE LT LU LV MT NL)
Observation	24 × 24	22 × 22	18 × 18
SER	0.722	0.621	0.571
R^2	0.895	0.924	0.918
Adjusted R^2	0.894	0.924	0.917
Border effect [exp. home]	4.411	5.989	15.661

Notes:
*, ** denote significance value of t-statistics at 0.01 and 0.05, respectively.
Results shown in this table are global. The analysis was performed with the use of country-aggregated data.

Source: own computation from Eurostat.

Euclidean distance are in line with the previous literature. When applying Euclidean distance to the more limited set of countries, our results are consistent with the findings of Chen (2004). Similarities between both studies include limiting the dataset to the manufacturing sector, introducing the same variables in the gravity model, excluding countries affected by the Rotterdam effect (although not explicitly mentioned in her publication), as well as applying a similar methodology to estimate intra- and international distance. Chen (2004, p. 98) indicated that European countries trade six times more within their borders than with other European countries. Our estimation, with updated data and a few more countries, is very similar (6.485). Similarly, we computed Wei's (1996) distance conceptualization for comparative purposes (not shown in the table), obtaining identical results (1.7, excluding only Malta, Cyprus and Luxembourg).

Nevertheless, results based on the Euclidean distance shown in Table 7.3 are slightly higher than those reported in Clark and van Wincoop (2001), whose estimations reduced the border effect in Europe to 1.27 for 1981–1997. Conversely, in Nitsch's (2000) study, which covered a wide selection of European countries and measured distance as a function of each country's area, his border effect estimation (7 to 10 times) is larger than ours (Nitsch, 2000, p. 1099).

The next step in our analysis was to compute the gravity model separately for each country, thus the different international trade behaviour of manufactured goods could be observed. Individual analysis per country was completed excluding all the countries in which the Rotterdam effect was identified. The national production of the origin country was excluded from the model since it becomes a constant variable. Table 7A.1 in the Appendix evidences that the global border effect masks a range of national home bias figures, for which some general spatial trends were identified. Although results vary with the distance metric used, in general central countries tend to have lower border effect values than peripheral ones. The latter is particularly true if we compute distance as travel time and GTC. Whilst most central countries show border effects below 20, peripheral countries tend to score higher than 40, with Portugal obtaining extremely high values and being the only country showing multicollinearity of the home and distance variables. A more detailed view evidences countries whose border effect value is more sensitive to the distance measurement. Some of these showed a high stability (for example, France and other central countries), but others (Italy and other peripheral countries) are very sensitive to the distance metrics. Finally, country level results shown in the Annex (Table 7A.1) are in line with those reported by Chen (2004) when using Euclidean distances.

7.4.2 Influence of the Border Effect on the Market Potential Indicator

The market potential indicator (Hansen, 1959) estimates the accessibility of location *i* based on a direct relationship with the size of destination and an inverse relationship with the distance or cost between origin *i* and destination *j*. It can be expressed as:

$$A_i = \sum\nolimits_{j=1}^{n} \frac{m_j}{t_{ij}^{\alpha}} \qquad (7.4)$$

where m_j denotes the mass at destination j (GDP, population), t_{ij} the impedance between locations i and j (distance, travel time or cost) and α is a distance decay parameter that needs to be calibrated in order to control for overestimation/underestimation of long-distance relationships.

However, the previous section evidenced that spatial interaction between countries decreases not only with distance but also due to national borders. All variables being equal (including distance), domestic trade would be higher than international trade.

The relevance of the border effect on accessibility measurements can be illustrated by comparing the market potential calculated with and without considering the border effect. Thus, the border effect (b) of the origin country (i) was introduced in equation (7.4) as a weighting factor for the mass at international destinations, as follows:

$$A_i = \sum\nolimits_{j=1}^{n} \frac{m_j * \frac{1}{b}}{t_{ij}^{\alpha}}; b = \begin{cases} b_i, i \neq j \\ 1, i = j \end{cases} \qquad (7.5)$$

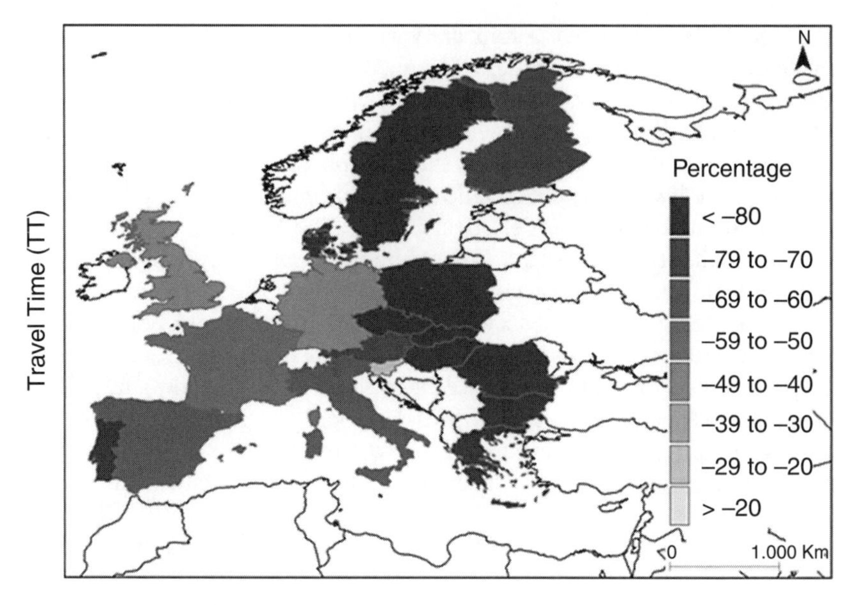

Source: own computation from Eurostat.

Figure 7.1 Difference in market potential after introducing border effect estimations (%)

We computed the market potential indicator for the selection of 18 European countries introducing the GDP as a proxy for the size of the destination country (m_j) and travel time (t) as the impedance variable. The value of the distance decay parameter (α) is the beta coefficient of the distance variable (travel time) in the global gravity model (β_5 in equation 7.1). The border effect (b) was computed individually for each origin country (i) and the results are shown in Table 7A.1 in the Appendix (model estimations for travel time).

Figure 7.1 shows the percentage difference of the market potential indicator after introducing the border effect. Negative values indicate that introducing the border effect lowers the value of the indicator. This reduction is greater in Eastern Europe and in particularly peripheral countries, like Portugal or Greece, than in Central Europe.

A more detailed analysis of the relative values of the market potential indicator (Figure 7.2) reveals that introducing the border effect (lower map) evidences a more polarized territory than what had been previously described (upper map). It is clear that international borders affect accessibility to a larger extent in peripheral countries than in Central Europe, increasing market potential disparities. Indeed, international

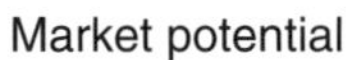

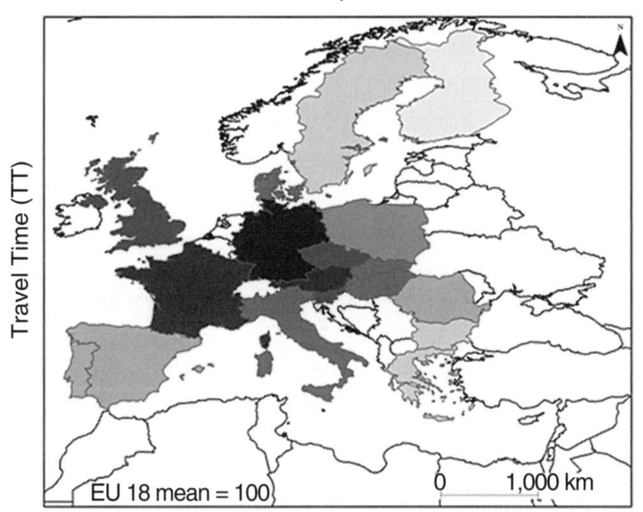

Market potential with border effect

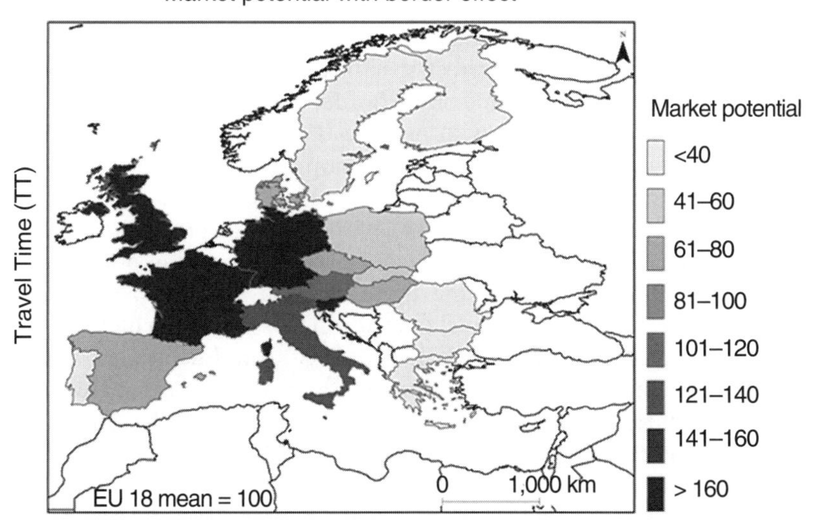

Source: own computation from Eurostat.

Figure 7.2 Market potential base 100 index

trade in some peripheral countries that joined the EU around 20 years ago (Greece, Portugal, Sweden, Finland) is still largely hampered by international borders in the European framework as evidenced by the results shown in Table 7A.1 in the Appendix.

7.5 CONCLUSIONS

Our main contribution has been to improve the estimation of the border effect as a first step towards improving the accuracy of accessibility measures when studying the international context. We briefly outlined the impact of introducing the border effect in accessibility analysis within the European Union. This leaves room for further research: for example, introducing trade with extra-EU countries into the analysis.

Previous studies using gravity models showed the sensitivity of border effects to different distance specifications. However the majority of studies have relied on Euclidean or great circle distances. In addition, many studies assumed that the economic activity of a country is concentrated in one point (usually the capital city). Both assumptions resulted in rough estimations of both national and international distances. Our study shows the importance of a more correct specification of the distance measure when estimating the magnitude of the border effect.

Our results are consistent with previous studies in the EU. Therefore, we agree with Anderson and van Wincoop's (2004, p. 733) claim that there is no hard evidence to refute the significance of the border effect, at least in the European framework. We agree with previous studies in that the choice of distance measurement plays a key role in border effect magnitude. Our results applying more realistic and complex distance measurements (particularly travel time and GTC) suggest an underestimation of previous home bias estimations derived from Euclidean distances. However, we cannot find a straightforward explanation for this trend that has been pointed out previously by other authors (Helliwell, 2002; Chen, 2004). This is certainly an interesting line of research that we leave open for future analysis.

Additionally, our results showed that the border effect is greater when countries affected by the Rotterdam effect are removed from the analysis. Overestimation of the ratio of exports to national production in those countries clearly contributes to a significantly reduced estimation of the overall European home bias, leading to unrealistic results.

Analysis at country level revealed higher border effects in peripheral EU countries. One explanation may be related to the fact that central countries are more competitive in international markets and are able to trade more outside their national borders, thus reducing their 'home bias'.

Border effects can be transferred to accessibility indicators, thereby improving estimations when analysing the international scene. We provide an example of how this could be done, resulting in a decrease of accessibility levels in the EU. Given the estimated size of the border effect across EU countries, accessibility was particularly reduced in peripheral countries.

As a policy conclusion, our border effect and market potential indicator results evidence the need for further policies to enhance territorial cohesion in order to facilitate international trade, particularly from peripheral countries.

We stress that our results only refer to manufactured products. Some studies undertaken at an industry level (Chen, 2004; Head and Mayer, 2002) found that bulky and heavy products (such as construction materials, wooden products, clay, and so on) have greater border effects. This is explained by the fact that such products travel shorter distances because they are harder and more costly to transport.

NOTES

* The views expressed are purely those of the authors and may not in any circumstances be regarded as stating an official position of the European Commission.

1. We decided to work at country level since official data on bilateral trade is only available at this level of analysis. There are some databases at NUTS 2 and NUTS 3 level for Europe, but these come from much-criticized estimates. Performing the analysis at a country level enabled us to compare our results with those obtained previously by other authors.
2. The Rotterdam (or Rotterdam/Antwerp) effect reveals that 'export figures from a country (A) to another country (B) may over-estimate the value of goods actually consumed in that country (B) if the importer forwards the goods on to another country (C)' (ONS, 2009, p. 195). It has been recognized as affecting European trade statistics (European Commission, 2006, p. 22).
3. NUTS 3 centroids were used for countries lacking NUTS 2 level (that is Estonia, Lithuania and Latvia). NUTS regions lacking connection with Europe mainland in TRANS-TOOLS' road and ferry network were removed from the analysis (that is Melilla, ES, and ultra-peripheral regions). Malta, Cyprus and Luxembourg were also removed from the analysis due to the inexistence of NUTS units below their national boundaries.
4. TRANS-TOOLS is a European transport network model used for EU policy analysis. Its main reference centre is the Joint Research Centre – Institute for Prospective Technological Studies (JRC-IPTS), which uses TRANS-TOOLS for impact assessments for different policy measures.

REFERENCES

Anderson, J.E. and E. van Wincoop (2001), 'Gravity with gravitas: a solution to the border puzzle', National Bureau of Economic Research.

Anderson, J.E. and E. van Wincoop (2004), 'Trade costs', *Journal of Economic Literature*, **42** (3), 691–751.
Chen, N. (2004), 'Intra-national versus international trade in the European Union: why do national borders matter?', *Journal of International Economics*, **63** (1), 93–118.
Clark, T.E. and E. van Wincoop (2001), 'Borders and business cycles', *Journal of International Economics*, **55** (1), 59–85.
European Commission (2006), *Statistics on the Trading of Goods – User Guide*, Luxembourg: Office for Official Publications of the European Communities.
European Union (2012), *EU Transport in Figures: Statistical Pocketbook 2012*, Luxembourg: Publications Office of the European Union.
Gutiérrez, J., A. Condeço-Melhorado, E. López and A. Monzón (2011), 'Evaluating the European added value of TEN-T projects: a methodological proposal based on spatial spillovers, accessibility and GIS', *Journal of Transport Geography*, **19** (4), 840–850.
Hansen, W.G. (1959), 'How accessibility shapes land use', *Journal of the American Institute of Planners*, **25** (2), 73–76.
Head, K. and T. Mayer (2002), 'Illusory border effects, distance mismeasurement inflates estimates of home bias in trade', Paris: Centre d'études prospectives et d'informations internationales.
Helliwell, J.F. (2002), 'Measuring the width of national borders', *Review of International Economics*, **10** (3), 517.
Helliwell, J.F. and G. Verdier (2001), 'Measuring internal trade distances: a new method applied to estimate provincial border effects in Canada', *Canadian Journal of Economics / Revue Canadienne D'Economique*, **34** (4), 1024–1041.
Hillberry, R. and D. Hummels (2003), 'Intranational home bias: some explanations', *Review of Economics and Statistics*, **85** (4), 1089–1092.
McCallum, J. (1995), 'National borders matter: Canada–US regional trade patterns', *American Economic Review*, **85** (3), 615–623.
Nitsch, V. (2000), 'National borders and international trade: evidence from the European Union', *Canadian Journal of Economics / Revue Canadienne D'Economique*, **33** (4), 1091–1105.
ONS (2009), *United Kingdom Balance of Payments: The Pink Book*, Southampton: Office for National Statistics.
Wei, S. (1996), 'Intra-national versus international trade: how stubborn are nations in global integration?', National Bureau of Economic Research.
Wolf, H.C. (1997), 'Patterns of intra- and inter-state trade', National Bureau of Economic Research.
Wolf, H.C. (2000), 'Intranational home bias in trade', *Review of Economics and Statistics*, **82** (4), 555–563.

APPENDIX

Table 7A.1 National border effect of the manufacturing sector, EU27 (excl. BE, CY, EE, IE, LT, LU, LV, MT, NL), 2009

Models	AT	BG	CZ	DE	DK	ES
Euclidean distance						
home	1.992**	0.572	2.551**	2.318*	1.555	2.141**
Ln Distance ij	−1.337*	−2.658*	−1.126**	−0.823*	−1.132**	−1.408*
Ln GDP j	0.612*	1.154*	0.621*	0.672*	0.853*	0.773*
Adj ij	0.269‘	−0.219‘	0.671	0.273	−0.083‘	0.543
S.E.R.	0.446	0.319	0.589	0.294	0.690	0.377
R^2	0.932	0.976	0.894	0.970	0.879	0.971
Border effect	7.329	1.772	12.822	10.155	4.736	8.505
Network distance						
home	2.158**	1.823*	2.591**	2.307*	2.028	2.500*
Ln Distance ij	−1.377*	−2.513*	−1.189*	−0.839*	−1.079**	−1.384*
Ln GDP j	0.614*	1.141*	0.627*	0.678*	0.856*	0.776*
Adj ij	0.262‘	0.150‘	0.582	0.248	−0.100‘	0.700
S.E.R.	0.422	0.344	0.568	0.284	0.690	0.402
R^2	0.939	0.972	0.902	0.973	0.879	0.968
Border effect	8.655	6.192	13.340	10.047	7.600	12.180
Travel time						
home	2.138**	2.296*	2.784*	2.329*	3.010***	3.131*
Ln Distance ij	−1.301*	−2.343*	−1.082*	−0.729*	−0.638	−1.084**
Ln GDP j	0.580*	1.078*	0.598*	0.687*	0.865*	0.805*
Adj ij	0.218‘	1.078**	0.679	0.265	0.119‘	1.007***
S.E.R.	0.402	3.079	0.559	0.285	0.809	0.430
R^2	0.945	0.970	0.905	0.972	0.834	0.963
Border effect	8.483	9.939	16.184	10.270	20.285	22.887
GTC						
home	2.949*	2.422***	2.671**	2.048*	2.003	3.151*
Ln Distance ij	−0.955*	−2.257*	−1.268**	−1.010*	−1.259**	−1.120**
Ln GDP j	0.652*	1.227*	0.706*	0.726*	0.911*	0.809*
Adj ij	0.922**	0.386‘	0.446	0.209	−0.134‘	1.090***
S.E.R.	0.479	0.660	0.603	0.271	0.704	0.947
R^2	0.921	0.897	0.889	0.975	0.874	0.959
Border effect	19.091	11.268	14.453	7.756	7.408	23.361

Notes:
*, **, ***, and ‘ denote significance value of t-statistics at 0.01, 0.05, 0.1 and 0.5, respectively
Observations: 1 × 18

Source: own computation from Eurostat.

Table 7A.1 (continued)

Models	FI	FR	GB	GR	HU
Euclidean distance					
home	1.957**	3.106*	2.913*	1.910	1.801***
Ln Distance ij	−1.402*	−0.285	−0.365	−1.676*	−1.106*
Ln GDP j	0.910*	0.886*	1.001*	0.924*	0.787*
Adj ij	0.783	0.329	0.000*	1.732**	0.414
S.E.R.	0.339	0.311	0.311	0.541	0.520
R^2	0.974	0.977	0.977	0.928	0.881
Border effect	7.079	22.324	18.406	6.752	6.056
Network distance					
home	3.024*	3.181*	2.920*	2.733**	2.198**
Ln Distance ij	−1.287*	−0.261	−0.379	−1.609*	−1.097*
Ln GDP j	0.907*	0.886*	1.000*	0.892*	0.773*
Adj ij	0.918***	0.349	0.000*	1.929**	0.415
S.E.R.	0.347	0.314	0.309	0.555	0.517
R^2	0.972	0.976	0.978	0.924	0.882
Border effect	20.571	24.066	18.544	15.375	9.006
Travel time					
home	2.192***	3.135*	3.333*	3.758*	2.276**
Ln Distance ij	−1.650*	−0.261	−0.099‘	−1.620*	−1.033*
Ln GDP j	0.930*	0.901*	1.047*	0.901*	0.722*
Adj ij	1.444*	0.306	0.000*	2.280*	0.396
S.E.R.	0.396	0.309	0.327	0.585	0.509
R^2	0.964	0.977	0.975	0.916	0.886
Border effect	8.953	22.990	28.021	42.842	9.735
GTC					
home	3.742*	3.188*	3.753*	3.833*	2.435*
Ln Distance ij	−1.139*	−0.302	0.240	−1.207***	−1.148*
Ln GDP j	0.921*	0.887*	1.108*	0.940*	0.833*
Adj ij	1.371*	0.377	0.000*	2.579**	0.381
S.E.R.	0.382	0.314	0.323	0.702	0.521
R^2	0.967	0.976	0.975	0.879	0.880
Border effect	42.165	24.252	42.654	46.187	11.420

IT	PL	PT	RO	SE	SI	SK
2.284*	2.507*	5.192*	1.897**	2.160*	−1.447	1.387
−1.301*	−1.332*	−0.285‘	−1.996*	−1.469*	−1.655*	−1.421*
0.781*	0.758*	1.170*	1.114*	0.997*	0.490*	0.810*
−0.243	0.409	1.596**	0.990**	1.125**	0.151‘	0.417
0.370	0.353	0.378	0.455	0.375	0.458	0.551
0.960	0.958	0.977	0.941	0.970	0.921	0.895
9.818	12.265	179.856	6.664	8.667	0.235	4.001
2.363*	2.598*	5.172*	2.452*	2.616*	0.153‘	1.861***
−1.319*	−1.312*	−0.349‘	−1.920*	−1.404*	−1.788*	−1.433*
0.777*	0.741*	1.165*	1.085*	0.990*	0.500*	0.794*
−0.220	0.363	1.548***	1.040**	1.169**	0.078‘	0.390
0.380	0.344	0.376	0.481	0.404	0.421	0.550
0.958	0.960	0.977	0.934	0.965	0.933	0.895
10.622	13.431	176.201	11.611	13.684	1.165	6.428
2.799*	2.689*	5.077*	2.410*	2.401*	0.349‘	2.039**
−1.002*	−1.274*	−0.392	−1.987*	−1.632*	−1.637*	−1.334*
0.744*	0.692*	1.168*	1.043*	0.952*	0.469*	0.744*
−0.128‘	0.346	1.536**	1.017**	2.042*	0.158‘	0.471
0.427	0.400	0.370	0.431	0.423	0.417	0.525
0.947	0.947	0.978	0.947	0.962	0.935	0.905
16.427	14.720	160.305	11.134	11.037	1.418	7.685
4.362*	2.659*	5.358*	3.373*	2.343*	1.527	2.338**
−1.561*	−1.374*	−0.302‘	−1.499*	−1.708*	−1.616*	−1.388*
0.774*	0.808*	1.177*	1.098*	1.057*	0.619*	0.854*
−0.202	0.282	1.658**	1.333***	1.055***	1.301**	0.447
0.378	0.385	0.378	0.682	0.463	0.613	0.617
0.958	0.951	0.977	0.868	0.954	0.859	0.869
78.434	14.288	212.403	29.165	10.411	4.604	10.364

8. Mapping transport disadvantages of elderly people in relation to access to bus stops: contribution of geographic information systems

Vitor Ribeiro, Paula Remoaldo and Javier Gutiérrez

8.1 INTRODUCTION

Increases in car ownership and improvements in highway networks have contributed to significant growth in the use of cars. In addition, land use planning also has promoted the dispersion of residential, service and business facilities throughout urban spaces. As a result, spatial dynamics contribute to the generation of spaces outside traditional urban centres, characterized by dependence on cars. Consequently, people who do not have or cannot drive cars may be excluded from services that are generally available.

Several authors have emphasized the link between transport limitations and social exclusion (Hine, 2000; Hine and Mitchell, 2001a; Knowles et al., 2008). Transport policies have a significant correlation with social exclusion (Hine, 2000) and this correlation begins with the accessibility of goods and services (Preston and Raje, 2007; Cebollada, 2008). In fact, social exclusion is not necessarily the result of a lack of opportunities; it may well be due to a lack of access to those opportunities (Kenyon et al., 2002; Preston and Raje, 2007). Access to the opportunities that are available is an individual's fundamental right, and the denial of access to those opportunities constitutes an exclusion factor.

The lack of public transport in some areas limits the extent to which people without cars can take advantage of such opportunities (Bavoux et al., 2005). Access to public transport by walking is particularly critical for vulnerable socio-demographic groups that are subjected to transport-related social exclusion, such as elderly people. Most elderly people rely on public transport. A range of barriers to walking can affect public transport access, such as the existence of stairs connecting two streets, security in the street, bridges, psychological factors, types of footwear or mobility constraints affecting the elderly. In addition, it is well known that elderly people walk more slowly than young people and other adults and that they are very sensitive to the effect of slope. The objective of this chapter is methodological: we integrate the street slope and the speed at which elderly people walk in an analysis of the accessibility of public transport using a geographic information system (GIS). These two vital factors have rarely been considered together in measuring the accessibility of public transport from the perspective of transport-related social exclusion. In this chapter the category of 'elderly' refers to people over 65 years old. The municipality of Braga in northwest Portugal is used as an example.

The remainder of the chapter is divided into six sections. Section 8.2 provides a literature review of both transport-related social exclusion and the analysis of the accessibility of public transport using GIS. Section 8.3 presents the main features of the area of study. Section 8.4 describes the data collected and the methodology used. Section 8.5 presents the main results and highlights the differences between the proposed methodology (including slopes and the speeds at which elderly people walk) and traditional methodologies (considering that all streets are flat and that walking speeds are standard and not group specific). Finally, section 8.6 presents our conclusions, based on the results of the study.

8.2 RESEARCH BACKGROUND

8.2.1 Understanding Transport-Related Social Exclusion

It is widely accepted that there has been a shift from the traditional approach, which considered that individuals and groups were excluded because they existed or operated outside a social norm (Hine and Mitchell, 2001b). Preston and Raje (2007) acknowledged that social exclusion was not synonymous with income-based deprivation, because it is possible for an individual to have a high-level income and still be socially excluded. As such, social exclusion can be seen as a multidimensional process with a dynamic profile, which individuals can move into and out of over time, and not only as the consequence of unemployment (Atkinson and Hills, 1998; Church et al., 2000; Lyons, 2003; Cebollada, 2008). McDonagh (2006) saw social exclusion as a cumulative marginalization process that was tied in with production, consumption, social networks, decision-making and access to an adequate quality of life.

It is thus recognized that poverty and social exclusion are not synonymous. Poverty can, indeed, lead to social exclusion and as a result is considered as one dimension of social exclusion (Pascal and Bourgeat, 2008). Poverty is more related to unequal access to material resources, while social exclusion, as a broader concept, focuses on unequal access to participation in society (Kenyon et al., 2002). Poverty, however, correlates negatively with the use of motorized transport (Salon and Gulyani, 2010).

Social exclusion implies that people or households cannot participate fully in society, not just because they are poor but also because of other factors, such as unemployment or the lack of access to facilities and services (Church et al., 2000). Kenyon et al. (2002) identified eight dimensions of social exclusion: economic, societal, social networks, organized political, personal political, living space, temporal and mobility. This multidimensional profile was also recognized by Chakravarty and D'Ambrosio (2006).

Concerning the study of the relationship between transport and social exclusion, Church et al. (2000) described two main approaches: the 'category approach' and the 'spatial approach'. The former focuses on travel patterns, attitudes and the needs of particular social groups; while the latter is more concerned with spatial accessibility. The spatial dimension of this relationship is dependent largely on the transport network system, since it is responsible for linking the locations of individuals and groups with the locations of opportunities, such as health or education facilities. Transport-related social exclusion refers to individuals who are physically unable to access the activities that are required to participate in society (Church et al., 2000; Preston and Raje, 2007; Kenyon, 2011). The lack of public transport in some areas limits people without cars from taking advantage of such opportunities (Bavoux et al., 2005). Frequently, important services are located in places that are difficult to reach without a car. These locations are rarely based on transport and accessibility criteria or on vulnerable areas identified by local authorities (Lucas, 2006). Increasing ownership of cars, combined with the decreasing accessibility of bus stops for those who must walk to them, has led to the growth of social exclusion areas (Lucas, 2006). Public transport thus has an important role in social exclusion policies, since it can be a major factor in mitigating social exclusion (Tyler, 2002). Poor public transport systems can lead to an increase in social exclusion (Stanley and Lucas, 2008).

Susceptibility to transport exclusion varies among different age groups. Low-income or unemployed individuals, children and young people, the elderly, women, the disabled, outer urban dwellers and ethnic minorities are identified as more susceptible groups (Dodson et al., 2006; Dodson et al., 2007). More work is needed in the field of accessibility and social inclusion (Farrington, 2007; Preston and Raje, 2007) and the lack of studies focusing on elderly people or others who must rely on public transport is particularly significant, in view of the vulnerability of these groups (Engels and Liu, 2011).

8.2.2 Accessibility to Bus Stops and GIS

Most users of public transport in urban areas access bus stops by walking. Accessibility to public transport facilities is perceived in spatial terms as the physical proximity to transit stops or stations. Coverage analyses have been used extensively to measure the accessibility of public-transport facilities. Access coverage is certainly important in public-transit planning, because this is the means by which the service is provided to riders (Wu and Murray, 2005). The aim of this measurement is to calculate the population or the number of employed people within a certain distance or time threshold with respect to one or several access points. Since the stops or stations on the transit network are represented by these points, the measure of accumulated opportunities provides an estimation of the numbers of people who have access to public transport networks because they live or work within walking distance of these networks (Gutiérrez and García-Palomares, 2008).

GIS has been used widely to assess the accessibility of public transport (Kwan et al., 2003), since it provides a flexible framework for estimating the population served (O'Neill et al., 1992; Murray et al., 1998). The usual practice for coverage analysis in a GIS is to delineate buffers (bands) around transit facilities based on Euclidean (straight-line) distance and to calculate the number of people who live within defined distance thresholds. This procedure (the buffer method) is the most commonly used method in transit-demand studies (Murray et al., 1998; O'Sullivan et al., 2000; Ryan and Getz, 2005; Oh and Jeong, 2007; Potoglou and Kanaroglou, 2008). However, the actual walking distance is longer than the Euclidean distance, because streets are generally not straight. As a result, the buffer (straight-line distance) method tends to overestimate the area and the population covered by the public transport network, leading to unrealistic results. In order to avoid the overestimation of the buffer method, it is possible to use the capabilities of GIS network analysis to calculate distances or travel times along a street network (network distance) (O'Neill et al., 1992; Hsiao et al., 1997; Horner, 2004). The key finding from network-based analysis was that it consistently returned a more conservative estimate of service area coverage than Euclidean access distances, irrespective of the transit feature considered or the population studied (Gutiérrez and García-Palomares, 2008).

Most studies have used standard distance thresholds in order to delimit the service area of transit facilities (bus stops or stations). In these studies, distance thresholds of 0.25 mile (400 metres) were used to determine access to bus stops and 0.50 mile (800 metres) for metro or railway stations (O'Neill et al., 1992; Hsiao et al., 1997; Murray, 2001; Kuby et al., 2004; Biba et al., 2010). Usually these distance thresholds are applied equally to all population groups and urban spaces, but accessibility is determined not only by place attributes but also by a person's characteristics (for example, age, wealth, health, gender) (Luo, 2004; Wang and Luo, 2005; Farrington, 2007). Studies regarding accessibility to public transport have demonstrated that walking distances may vary according to different types of socio-demographic groups. Thus, for example, young people walk greater distances than older people (El-Geneidy et al., 2011; García-Palomares et al., 2013). In addition a number of studies have shown that elderly people walk more slowly than younger people (Himann et al., 1988; Bendall et al., 1989; Bohannon, 1997; Steffen et al., 2002; Willis et al., 2004; Kang and Dingwell, 2008; Lindemann et al., 2008).

Gradients are important factors in many cities, because they affect walking speed and travel time. The impact of gradients on the estimation of walking time has been based largely on the 'Naismith rule', which was developed in the early nineteenth century for hilly terrain (Chiou et al., 2010). Most studies that have considered the impact of slope on travel time were related to hill walkers or mountaineers (Rees, 2004; Chiou et al., 2010), but the environment of an urban street is different from the conditions that exist in the mountains. Recently Finnis and Walton (2008) produced one of the few urban studies that have linked walking speed with slope, and they showed how average walking speeds varied depending on whether people were walking up or down the slopes. The authors measured pedestrian walking speeds in four New Zealand cities (Auckland, Wellington, Palmerston North and Levin). A total of 1847 pedestrians' movements were recorded using a video camera on 13 sites. They recorded a 5 metre section on the walkway in cases where it sloped.

Despite these findings, walking accessibility studies based on GIS analysis have seldom differentiated specific speeds of socio-demographic groups and the characteristics of the slope of the street. One exception was the research conducted by Colclough and Owens (2010), in which the authors used the results provided by Finnis and Walton (2008) and Willis et al. (2004) to evaluate accessibility levels in west Northamptonshire, UK, in a GIS environment. However, they assumed that the speeds of all the socio-demographic groups had the same rate of change in response to the gradients, which ignored the fact that the slope effect is particularly important for elderly people. These authors concluded that specific localized areas that have significant gradients influenced the results, but the consistent variation in walking speed due to demographic type had a greater effect on the results.

8.3 CASE STUDY

The study area was the municipality of Braga, which is situated in northwest Portugal (Figure 8.1). Braga (181 494 inhabitants in 2011) is the third-largest city in Portugal, after the metropolitan areas of Lisbon and Porto. Throughout the last decade Braga has experienced high urban dynamics and there has been a very significant growth in infrastructure. The growth rate of the population in the last inter-Census period (2001–2011) was 11 per cent, one of the highest in northern Portugal. The proportion of older people grew from 11 per cent in 2001 to 13 per cent in 2011, while the percentage of young people decreased from 19 per cent to 16 per cent (Table 8.1).

Braga still preserves a compact urban form. The main economic activities (health, education, administration and tourism) are concentrated within or near the central business district (CBD) but in the last few years the dispersion of new urban development (residential areas, shopping centres and industrial areas) has generated complex flows. Private cars continue to be the main transport mode outside the city centre, because the public urban transport system has difficulty meeting demand.

The public transport network has 76 bus lanes and about 1590 bus stops. Lately there has been an increase in complaints from passengers and the general population concerning the availability of public transport and the locations of bus stops. Since Braga is a very hilly municipality, slopes have a significant influence on the accessibility of bus stops. Among the city's streets, 67 per cent have gentle slopes in the range 0 per cent to 4 per cent; 24 per cent have moderate slopes in the range 4 per cent to 8 per cent; and 9 per cent have steep slopes (greater than 8 per cent).

8.4 DATA AND METHODS

8.4.1 Disaggregation of Spatial Demographic Data

Braga is divided into 62 parishes. The Census operation divides these parishes into small territorial units (213 sections and 1481 subsections in Braga). Population data for the elderly from the 2001 Census (the most recent with statistical subsection data) were provided by the National Statistics Institute of Portugal. Since the statistical subsection scale is not

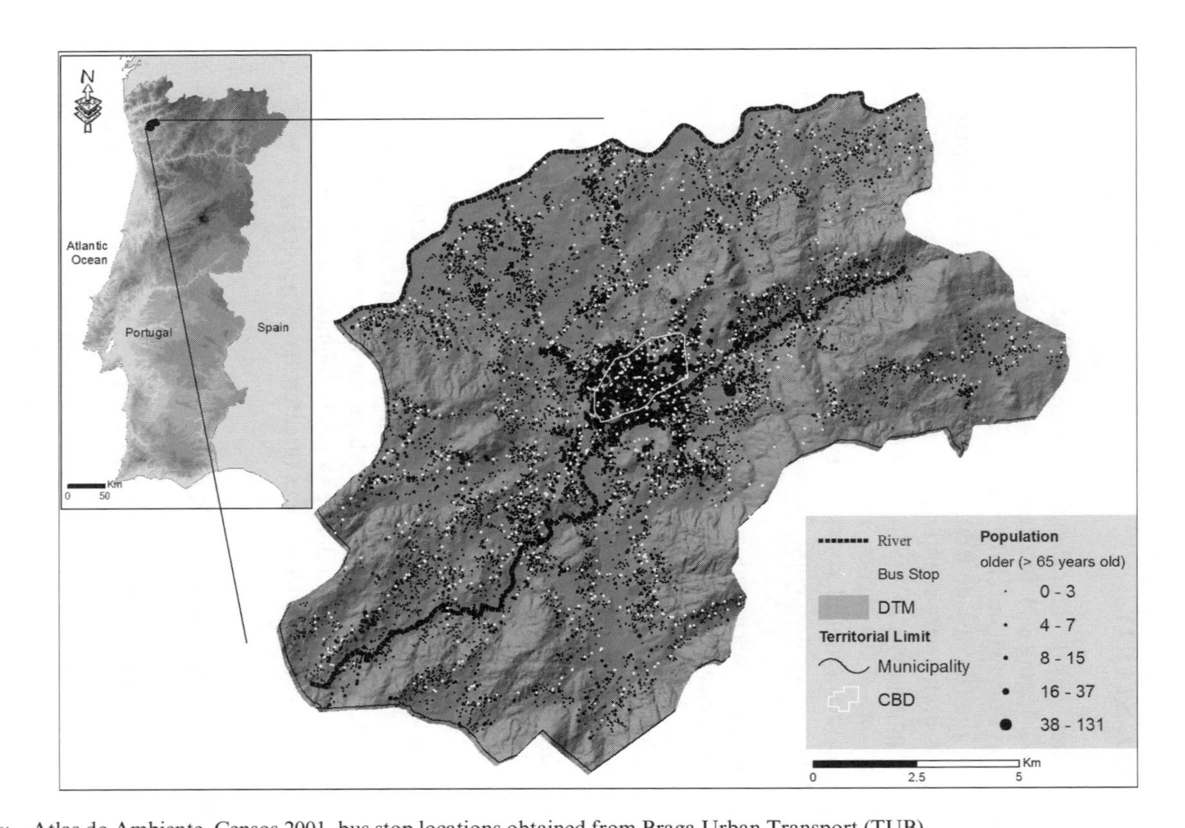

Sources: Atlas do Ambiente, Censos 2001, bus stop locations obtained from Braga Urban Transport (TUB).

Figure 8.1 Area of study: elderly population, bus stops and Digital Terrain Model (DTM)

Table 8.1 Population age structure by major age groups, 2001 and 2011, in Braga

Year	2001		2011	
Age group	Abs.	%	Abs.	%
0–14 years old	31 291	19	29 928	16
15–64 years old	115 703	70	128 654	71
65+ years old	17 893	11	23 247	13
Total	164 886	100	181 829	100

Source: Pordata.

compatible with the analysis of the accessibility of urban transport bus stops, dasymetric mapping was used to obtain more disaggregated data related to individual and household characteristics. In this technique we used additional or ancillary data to refine the population's locations and mask inhabited areas, an approach that the literature often refers to as 'filtered areal weighting'. Several methods can be used in dasymetric mapping. The traditional, binary-mask method considers the classes of ancillary data as either populated or unpopulated (Eicher and Brewer, 2001; Mennis and Hultgren, 2006). The limiting variable dasymetric is another method, which uses known densities to balance the population surface between different land-use classes (Tapp, 2010). Recent studies have commonly used land-use data, for example, CORINE land cover (a map of the European environmental landscape based on interpretation of satellite images) (Langford and Unwin, 1994; Langford, 2006; Mennis and Hultgren, 2006; Langford, 2007; Su et al., 2010; Gallego et al., 2011). Maantay and Maroko (2009) stated that this type of dasymetric disaggregation can be of limited utility in urban areas. The authors used cadastral-based data as the ancillary data.

In this research, since a cadastral-based database was not available, a database was created based on satellite images in order to identify each residential building within the municipality by a polygon feature. Once this residential database was created, the population of each subsection was disaggregated according to the formula:

$$p_b = p_s \frac{a_b}{a_s} \tag{8.1}$$

where:

p_b = population in building b

p_s = population in subsection s
a_b = area of residential building b
a_s = residential area of subsection s.

After this spatial disaggregation of population, 7223 residential building centroids were obtained to perform accessibility measures. A shift from an average of 111 individuals by subsection to 22 individuals at each centroid was obtained.

8.4.2 Network of Streets

A network containing all the streets and roads of Braga was used in order to calculate the accessibility of bus stops. In this network, information for length, slope, speed and the specific walking time of each arc was stored. In order to calculate the slope of the arc, as Colclough and Owens (2010) did, every arc greater than 100 metres was split to incorporate different slope levels along the arcs and x, y coordinates were calculated for the nodes defined by the beginning and end of each arc. Those fields were used to create two-point file features (startpoint and endpoint) for which the elevation attributes were extracted from a surface grid. This information was then added to the street table as the start node and the end node fields, which were populated with elevation data. Based on this information, the slope of each arc was calculated using the formula:

$$g_a = \frac{h_a}{l_a}100 \tag{8.2}$$

where g_a is the slope of arc a in percentage, h_a is the absolute height difference between the start node and the end node of arc a and l_a is the length of the arc. The value of g_a can be positive or negative, indicating whether a pedestrian would have to walk downhill or uphill from the start node to the end node.

Walking times for elderly people were calculated in metres per minute, using different walking speeds according to the street's slope. The speeds for the different slopes (Table 8.2) were obtained from fieldwork, based on the following regularities:

1. At street sections with very gentle slopes (between 0 per cent and 2 per cent), the average walking speed of elderly people (50 m/min) was much lower than the average walking speed (about 80 m/min).
2. Elderly people are more sensitive to slopes than other adults.
3. On ascending sections, walking speed decreased as the slope increased. This effect was almost double for elderly people compared with adults.

Table 8.2 Walking speed of elderly people (m/min) as a function of percentage slope

% slope	Walking up		Walking down	
	Speed (m/min)	Slope Impact (%)	Speed (m/min)	Slope impact (%)
0–2	50	0	50	0
2–4	45	−10	50	0
4–6	40	−20	45	−10
6–8	35	−30	40	−20
8–10	30	−40	35	−30
>10	25	−50	30	−40

4. On descending sections, the relationship between slope and walking speed was more complex. Adults' walking speed increased with very gentle slopes and decreased on steep slopes. Elderly people felt unsafe on steep slopes and the fear of falling affected their walking speed.

The walking time of elderly people was clocked on different streets in 150 metre sections. Only individuals that completed the 150 metre route were considered. Apparently elderly people were anonymously followed and their walking speed was calculated in metres per minute.

8.4.3 Scenarios and Accessibility Calculation

Three scenarios were considered in order to analyse the accessibility of public transport for elderly people. The first scenario ignored the street gradients and assumed a standard walking speed (80 m/min). The second scenario considered the specific walking speed of elderly people (50 m/min) and ignored the street gradients. Finally the third scenario took into account the specific walking speeds of elderly people as a function of the street slopes (Table 8.2). Comparisons between these three scenarios allow us to demonstrate the effect of a specific walking speed and slope on the accessibility of public transport for elderly people.

Accessibility to bus stops was based on calculations of walking times along a network of streets. GIS network analysis tools allowed us to calculate the fastest walking routes from the households (building centroids) to the bus stops using different walking speeds. Once routes with the minimal walking times had been calculated, the average walking times to the nearest stop and the distribution of the elderly population according to isochrones were obtained. In order to analyse spatial differences in the effects of slope, average walking times were calculated according to parishes.

8.5 RESULTS

This section shows the results of the accessibility analysis according to the three scenarios mentioned above. Differences both in average walking time to the nearest stop and in the numbers of elderly people involved according to isochrones showed the effects of ignoring the specific walking speeds of elderly people and ignoring slopes when calculating the accessibility of public transport for elderly people.

In scenario 1 (average walking speed of 80 m/min), 69 per cent of the elderly population lived less than three minutes from a bus stop, which suggests that a large proportion of this vulnerable social group had excellent access to public transport. In scenarios 2 and 3, however (specific speed of elderly people without and with slope effect), the percentages of elderly people within that walking-time threshold were 46 per cent and 40 per cent respectively (Table 8.2). In contrast, the elderly population living more than five minutes from the bus stops (potentially excluded population) increased from 12.3 per cent in the first scenario to 28.5 per cent in the second and third scenarios. This demonstrated that ignoring the specific walking speeds of elderly people and ignoring the slope produced overestimates of the population covered by the first time band and underestimates of the potentially excluded population (Table 8.3).

This fact is also reflected in the average walking time to the nearest bus stop, which increased from 2.7 minutes in the first scenario to 4.3 in the second and 5.1 in the third (Table 8.3). Therefore, scenarios 1 and 2 resulted in significant underestimation of the average walking times required to access the bus stops (−47.2 per cent and −16.0 per cent, respectively) (Table 8.4). Elderly people walk at a lower speed than other adults and need more time to cover a given distance due to their physical condition and their fear of falling. Indeed falls are responsible for many immobility conditions and deaths among this socio-demographic group. As a result, scenarios 1 and 2 overestimate the number of elderly living within the walking-time threshold of three minutes and underestimate those living within the other walking-time intervals (Table 8.4).

The results shown in Tables 8.2 and 8.3 indicate that the slope effect in Braga is not as important as the specific walking speed of elderly people. The slope effect was responsible for a variation of −16.0 per cent in access time to bus stops and walking speed was responsible for a variation of −31.2 per cent (−47.2 per cent overall). However, the slope effect on the

Table 8.3 Accessibility of bus stops: number of elderly people covered and average walking time to the nearest bus stop by walking-time threshold

Walking-time thresholds (min)	Scenario 1: Standard walking speed ignoring slopes (80 m/min)		Scenario 2: Specific walking speed of elderly people ignoring slopes (50 m/min)		Scenario 3: Specific walking speed of elderly people accounting for slopes	
	Number of individuals	%	Number of individuals	%	Number of individuals	%
0–3	12 129	69.2	8 122	46.4	7 019	39.9
3–5	3 227	18.4	4412	25.2	5 575	31.7
5–10	1 804	10.3	3 607	20.6	3 704	21.1
10–15	254	1.4	924	5.3	876	5.0
>15	104	0.6	453	2.6	421	2.4
Total	17 518	100	17 518	100	17 595	100
Average walking time	Minutes 2.7	–	Minutes 4.3	–	Minutes 5.1	

Table 8.4 Overestimation and underestimation in scenarios 2 and 3 (see Table 8.3)

Walking-time thresholds (min)	Difference between scenarios 1 and 3		Difference between scenarios 2 and 3	
	Number of individuals	%	Number of individuals	%
0–3	4833	66.2	826	11.3
3–5	−828	−20.4	−1185	−29.2
5–10	−2377	−56.9	−1803	−43.1
10–15	−951	−78.9	−670	−55.6
>15	−678	−86.7	−349	−44.6
Average walking time	Minutes −2.4	% −47.2	Minutes −0.8	% −16.0

accessibility of public transport for elderly people exhibited a high spatial variation (Figure 8.2). Ignoring slope produces an average underestimate of 16.0 per cent in the walking time to the nearest bus stop, but underestimates are above average in hilly districts and above average in flat ones. The city centre, which has a large concentration of elderly people, presents

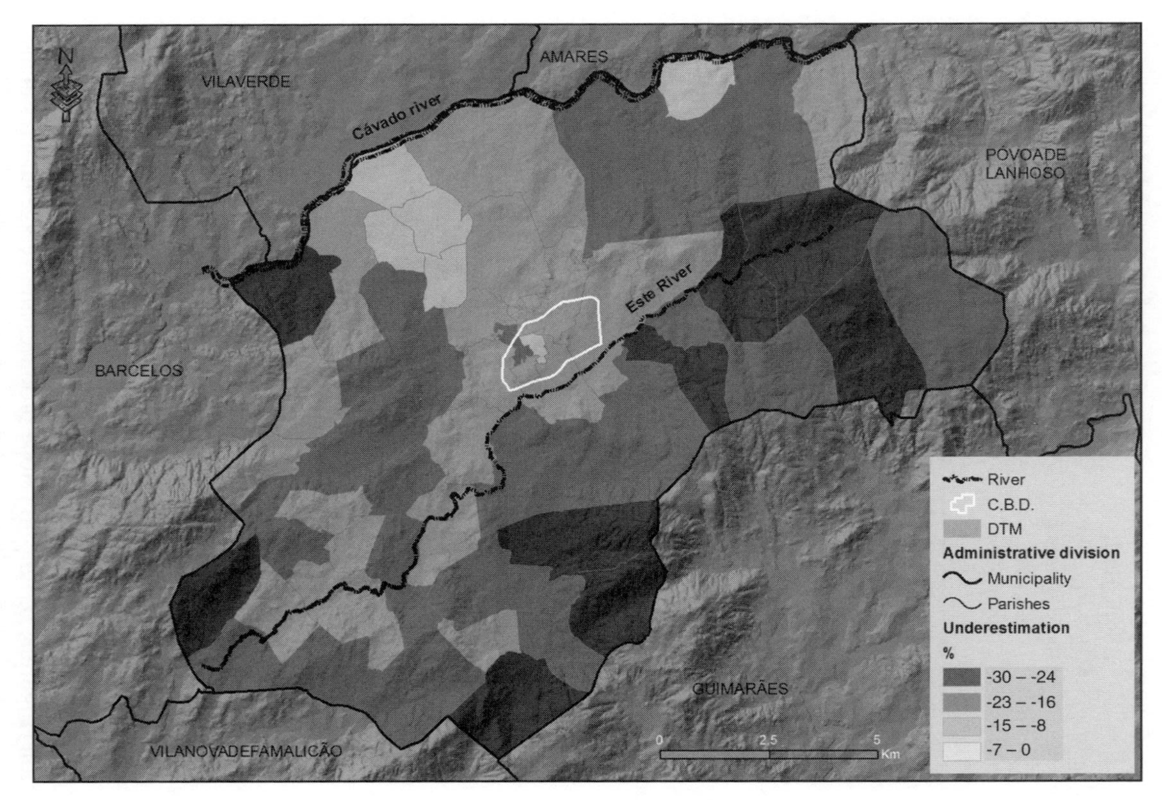

Figure 8.2 *Slope effect on the walking time of elderly people to bus stops according to parishes in Braga: differences (%) between scenarios 2 and 3*

underestimates that are below the average, while some peripheral districts had underestimates of about 30 per cent, so that in those districts the effect of ignoring the influence of slope was as important as the effect of ignoring the specific walking speed of elderly people.

8.6 FINAL REMARKS

Accessibility to public transport is essential because it enables people to participate in the opportunities offered by cities. This is particularly critical for vulnerable socio-demographic groups that are subjected to transport-related social exclusion, such as elderly people, who must rely heavily on public transport and who sometimes have difficulty walking. Most previous studies considered accessibility to public transport without taking into account the special characteristics of elderly people. As this population group walks more slowly than the average and is more sensitive to the effect of slope, it is necessary to consider these two factors to analyse the accessibility of public transport for elderly people.

The results reported in this chapter show that considering the specific walking speeds of elderly people and the slope effect provides more accurate results. The traditional methodology (standard walking speed without considering the slope effect) results in significant overestimation of the elderly population with good access to bus stops and it underestimates the potentially excluded population and the average time required to access bus stops. In the case of Braga, the effect of the specific walking speed of elderly people was more important than the slope effect. But the latter has a pronounced spatial variability, so that in hilly parishes the slope effect was as important as the effect of the specific walking speed of elderly people.

These results have important implications for transport planning and policies, particularly from the social point of view. The approach proposed in this chapter produces more accurate estimates than traditional methods. As a result, transport-related social exclusion can be more precisely measured and spatial inequalities are likely to be better understood. In addition, public transport networks can be designed considering the particular characteristics of elderly people, especially in hilly districts or in districts that have a high proportion of elderly people.

ACKNOWLEDGMENT

We would like to express our gratitude to the Portuguese Foundation for Science and Technology for the PhD grant to Vitor Ribeiro (Grant No. SFRH/BD/38762/2007).

REFERENCES

Atkinson, A. and J. Hills (1998), *Exclusion, Employment and Opportunity*, London: Centre for Analysis of Social Exclusion, London School of Economics.

Bavoux, J.J., F. Beaucire, L. Chapelon and P. Zembri (2005). *Géographie des transports*, Paris: Armand Colin.

Bendall, M.J., E.J. Bassey and M.B. Pearson (1989), 'Factors affecting walking speed of elderly people', *Age and Ageing*, **18** (5), 327–332.

Biba, S., K. Curtin and G. Manca (2010), 'A new method for determining the population with walking access to transit', *International Journal of Geographical Information Science*, **24** (3), 347–364.

Bohannon, R. (1997), 'Comfortable and maximum walking speed of adults aged 20–79 years: reference values and determinants', *Age and Ageing*, **26** (1), 15–19.

Cebollada, À. (2008), 'Mobility and labour market exclusion in the Barcelona Metropolitan Region', *Journal of Transport Geography*, **17** (3), 226–233.

Chakravarty, S.R. and C. D'Ambrosio (2006), 'The measurement of social exclusion', *Review of Income and Wealth*, **52** (3), 377–398.

Chiou, C.-R., W.-L. Tsai and Y.-F. Leung (2010), 'A GIS-dynamic segmentation approach to planning travel routes on forest trail networks in Central Taiwan', *Landscape and Urban Planning*, **97** (4), 221–228.

Church, A., M. Frost and K. Sullivan (2000), 'Transport and social exclusion in London', *Transport Policy*, **7** (3), 195–205.

Colclough, J.G. and E. Owens (2010), 'Mapping pedestrian journey times using a network-based GIS model', *Journal of Maps*, **6** (1), 230–239.

Dodson, J., N. Buchanan, B. Gleeson and N. Sipe (2006), 'Investigating the social dimensions of transport disadvantage I: towards new concepts and methods', *Urban Policy and Research*, **24** (4), 433–453.

Dodson, J., B. Gleeson, R. Evans and N. Sipe (2007), 'Investigating the social dimensions of transport disadvantage II: from concepts to methods through an empirical case study', *Urban Policy and Research*, **25** (1), 63–89.

Eicher, C.L. and C.A. Brewer (2001), 'Dasymetric mapping and areal interpolation: implementation and evaluation', *Cartography and Geographic Information Science*, **28** (2), 125–138.

El-Geneidy, A., A. Cerdá, R. Fischler and N. Luka, (2011), 'Using accessibility measures: a test case in Montréal evaluating the impacts of transportation plans', *Canadian Journal of Urban Research: Canadian Planning and Policy*, **20** (1), 81–104.

Engels, B. and G.-J. Liu (2011), 'Social exclusion, location and transport disadvantage amongst non-driving seniors in a Melbourne municipality, Australia', *Journal of Transport Geography*, **19** (4), 984–996.

Farrington, J.H. (2007), 'The new narrative of accessibility: its potential contribution to discourses in (transport) geography', *Journal of Transport Geography*, **15** (5), 319–330.
Finnis, K. and D. Walton (2008), 'Field observations to determine the influence of population size, location and individual factors on pedestrian walking speeds', *Ergonomics*, **51** (6), 827–842.
Gallego, F., F. Batista, C. Rocha and S. Mubareka (2011), 'Disaggregating population density of the European Union with CORINE land cover', *International Journal of Geographical Information Science*, **25** (12), 2051–2069.
García-Palomares, J.C., J. Gutiérrez and O.D. Cardozo (2013), 'Walking accessibility to public transport: an analysis based on microdata and GIS', *Environment and Planning B: Planning and Design*, **40** (6), 1087–1102.
Gutiérrez, J. and J.C. García-Palomares (2008), 'Distance-measure impacts on the calculation of transport service areas using GIS', *Environment and Planning B: Planning and Design*, **35** (3), 480–503.
Himann, J., D. Cunningham, P. Rechnitzer and D. Paterson (1988), 'Age-related changes in speed of walking', *Medicine & Science in Sports and Exercise*, **20** (2), 161–166.
Hine, J. (2000), 'Integration, integration, integration . . . Planning for sustainable and integrated transport systems in the new millennium', *Transport Policy*, **7** (3), 175–177.
Hine, J. and F. Mitchell (2001a), 'Better for everyone? Travel experiences and transport exclusion', *Urban Studies*, **38** (2), 319–332.
Hine, J. and F. Mitchell (2001b), *The Role of Transport in Social Exclusion in Urban Scotland*, Edinburgh: Scottish Executive Central Research Unit.
Horner, M. (2004), 'Spatial dimensions of urban commuting: a review of major issues and their implications for future geographic research', *Professional Geographer*, **56** (2), 160–173.
Hsiao, S., J. Lu, J. Sterling and M. Weatherford (1997), 'Use of geographic information system for analysis of transit pedestrian access', *Transportation Research Record*, **1604** (1), 50–59.
Kang, H.G. and J.B. Dingwell (2008), 'Separating the effects of age and walking speed on gait variability', *Gait and Posture*, **27** (4), 572–577.
Kenyon, S. (2011), 'Transport and social exclusion: access to higher education in the UK policy context', *Journal of Transport Geography*, **19** (4), 763–771.
Kenyon, S., G. Lyons and J. Rafferty (2002), 'Transport and social exclusion: investigating the possibility of promoting inclusion through virtual mobility', *Journal of Transport Geography*, **10** (3), 207–219.
Knowles, R., J. Shaw and I. Docherty (2008), *Transport Geographies – Mobilities, Flows and Spaces*, Malden: Blackwell.
Kuby, M., A. Barranda and C. Upchurch (2004), 'Factors influencing light-rail station boardings in the United States', *Transportation Research Part A: Policy and Practice*, **38** (3), 223–247.
Kwan, M.-P., A.T. Murray, M.E. O'Kelly and M. Tiefelsdorf (2003), 'Recent advances in accessibility research: representation, methodology and applications', *Journal of Geographical Systems*, **5** (1), 129.
Langford, M. (2006), 'Obtaining population estimates in non-census reporting zones: an evaluation of the 3-class dasymetric method', *Computers, Environment and Urban Systems*, **30** (2), 161–180.
Langford, M. (2007), 'Rapid facilitation of dasymetric-based population

interpolation by means of raster pixel maps', *Computers, Environment and Urban Systems*, **31** (1), 19–32.

Langford, M. and D.J. Unwin (1994), 'Generating and mapping population density surfaces within a geographical information system', *Cartographic Journal*, **31** (1), 21–26.

Lindemann, U., B. Najafi, W. Zijlstra, K. Hauer, R. Muche, C. Becker and K. Aminian (2008), 'Distance to achieve steady state walking speed in frail elderly persons', *Gait and Posture*, **27** (1), 91–96.

Lucas, K. (2006), 'Providing transport for social inclusion within a framework for environmental justice in the UK', *Transportation Research Part A: Policy and Practice*, **40** (10), 801–809.

Luo, W. (2004), 'Using a GIS-based floating catchment method to assess areas with shortage of physicians', *Health and Place*, **10** (1), 1–11.

Lyons, G. (2003), 'The introduction of social exclusion into the field of travel behaviour', *Transport Policy*, **10** (4), 339–342.

Maantay, J. and A. Maroko (2009), 'Mapping urban risk: flood hazards, race, and environmental justice in New York', *Applied Geography*, **29** (1), 111–124.

McDonagh, J. (2006), 'Transport policy instruments and transport-related social exclusion in rural Republic of Ireland', *Journal of Transport Geography*, **14** (5), 355–366.

Mennis, J. and T. Hultgren (2006), 'Intelligent dasymetric mapping and its application to areal interpolation', *Cartography and Geographic Information Science*, **33** (3), 179–194.

Murray, A.T. (2001), 'Strategic analysis of public transport coverage', *Socio-Economic Planning Sciences*, **35** (3), 175–188.

Murray, A.T., R. Davis, R.J. Stimson and L. Ferreira (1998), 'Public transportation access', *Transportation Research Part D: Transport and Environment*, **3** (5), 319–328.

O'Neill, W.A., R.D. Ramsey and J.C. Chou (1992), 'Analysis of transit service areas using geographic information systems', *Transportation Research Record*, **1364**, 131–138.

O'Sullivan, D., A. Morrison and J. Shearer (2000), 'Using desktop GIS for the investigation of accessibility by public transport: an isochrone approach', *International Journal of Geographical Information Science*, **14** (1), 85–104.

Oh, K. and S. Jeong (2007), 'Assessing the spatial distribution of urban parks using GIS', *Landscape and Urban Planning*, **82** (1–2), 25–32.

Pascal, B. and S. Bourgeat (2008), *Dictionnaire de géographie, 4e édition*, Paris: Hartier.

Potoglou, D. and P.S. Kanaroglou (2008), 'Modelling car ownership in urban areas: a case study of Hamilton, Canada', *Journal of Transport Geography*, **16** (1), 42–54.

Preston, J. and F. Raje (2007), 'Accessibility, mobility and transport-related social exclusion', *Journal of Transport Geography*, **15** (3), 151–160.

Rees, W.G. (2004), 'Least-cost paths in mountainous terrain', *Computers and Geosciences*, **30** (3), 203–209.

Ryan, S.J. and W.M. Getz (2005), 'A spatial location-allocation GIS framework for managing water sources in a savanna nature reserve', *South African Journal of Wildlife Research*, **35** (2), 163–178.

Salon, D. and S. Gulyani (2010), 'Mobility, poverty, and gender: travel "choices" of slum residents in Nairobi, Kenya', *Transport Reviews*, **30** (5), 641–657.

Stanley, J. and K. Lucas (2008), 'Social exclusion: what can public transport offer?', *Research in Transportation Economics*, **22** (1), 36–40.
Steffen, T.M., T.A. Hacker and L. Mollinger (2002), 'Age- and gender-related test performance in community-dwelling elderly people: six-minute walk test, berg balance scale, timed Up & Go test, and gait speeds', *Physical Therapy*, **82** (2), 128–137.
Su, M.-D., M.-C. Lin, H.-I. Hsieh, B.-W. Tsai and C.-H. Lin (2010), 'Multi-layer multi-class dasymetric mapping to estimate population distribution', *Science of the Total Environment*, **408** (20), 4807–4816.
Tapp, A.F. (2010), 'Areal interpolation and dasymetric mapping methods using local ancillary data sources', *Cartography and Geographic Information Science*, **37** (3), 215–228.
Tyler, N. (2002), *Accessibility and the Bus System: From Concepts to Practice*, London: Thomas Telford.
Wang, F. and W. Luo (2005), 'Assessing spatial and nonspatial factors for health-care access: towards an integrated approach to defining health professional shortage areas', *Health and Place*, **11** (2), 131–146.
Willis, A., N. Gjersoe, C. Havard, J. Kerridge and R. Kukla (2004), 'Human movement behaviour in urban spaces: implications for the design and modelling of effective pedestrian environments', *Environment and Planning B: Planning and Design*, **31** (6), 805–828.
Wu, C.S. and A.T. Murray (2005), 'Optimizing public transit quality and system access: the multiple-route, maximal covering/shortest-path problem', *Environment and Planning B: Planning and Design*, **32** (2), 163–178.

PART III

Accessibility as a driver of spatial interaction

9. Productivity and accessibility of road transportation infrastructure in Spain: a spatial econometric approach

Pelayo Arbués, Matías Mayor and José Baños

9.1 INTRODUCTION

Measuring the economic effects of public infrastructure improvements on the productivity of private capital has been at the centre of the academic debate for the last two decades. The concept underlying these papers is that public capital plays a significant role as an input factor in the production process. The first empirical studies on this issue appeared in the 1970s (Mera, 1973), but it was only with the extraordinary results obtained by Aschauer (1989) that the research community showed a revived interest in the effects of public infrastructure improvements. In these early works the authors found that public capital exerted a large and significant effect on output. Aschauer estimated an output elasticity of approximately 0.4 and the results in the study by Munnell and Cook (1990) ranged from 0.31 to 0.39. In an era when the productivity growth of most Organisation for Economic Co-operation and Development (OECD) countries experienced a significant slowdown, policy administrators and scholars wondered whether this might be caused at least in part by insufficient public capital. In this context Aschauer's main findings were appealing as an increase in public investment in infrastructure seemed a straightforward solution to an alarming slowdown in productivity.

Subsequent research failed to find significant positive effects of public capital on private output (Holtz-Eakin and Schwartz, 1995). Regional data were used notably in the articles with models that obtained low estimates of marginal productivity, while the early studies obtained large effects with national data. Differences in estimates of public capital also point to a geographical pattern. Certain researchers suspected the existence of spillover effects (Cohen and Morrison, 2004) which are also known as leakages.

These spillover effects indicate that the effects generated from public capital investment would not be confined to the region in which the infrastructure is located. If spillovers were present, part of the effect of public capital would be underestimated by using regional data. The development of spatial econometric methodologies has allowed an exploration of the potential existence of spatial interdependencies among geographical units.

While there is little doubt that enterprises need a minimum level of public infrastructure to generate output to sell in markets, it should not be expected that the marginal output effect of extra public infrastructure remains constant at every level. In the case of road transportation, building one interstate network might cause a significant increase in productivity but building a second might not (Hulten, 2004). Considered in terms of productivity changes, the significance of the role that transportation infrastructures play in the economy of a region is determined by the services it provides. Improvements in these services are expected to reduce generalized transportation costs as a result of shorter distances, less congestion and higher speeds that reduce fuel, capital and labour costs (Forkenbrock and Foster, 1990). However, transportation projects create other significant spatial location services in addition to reducing travel and logistics costs. They may enlarge the market potential of businesses by enabling them to serve broader markets more economically. In addition, improvements in the transportation system can provide firms with a greater variety of specialized labour skills and input products, making them more productive. Rietveld (1994) offers a description of the spatial development effects resulting from transportation infrastructure supply as a complete theoretical framework. Measuring infrastructure as a stock fails to account for the actual supply of the services that determine its contributions to productivity (Oosterhaven and Knaap, 2003). To overcome this problem some authors have proposed the use of accessibility measures instead of the stock of infrastructure (Forslund and Johansson, 1995; Rietveld and Nijkamp, 2000).

Road transportation infrastructure projects in Spain have been promoted through the implementation of the Infrastructure and Transport Strategic Plan that raised the quality of Spain's road transportation network to European standards in a short period of time. The objective of this study is to measure the output effect of road transportation infrastructure in Spain in the period between 1997 and 2006. In particular we estimate a production function using a panel dataset of Spanish provinces in order to account for marginal productivity effects within a province and to document the existence of spillover effects outside the provincial boundaries through the use of spatial econometric methodologies. Road infrastructure endowment indicators are measured using three different indicators: an accessibility measure, the traditional road stock indicators and a variable to accommodate the stock indicator to the degree of utilization. This framework attempts to overcome the usual problems in the literature that may be causing the current ambiguous state of results as pointed out by Mikelbank and Jackson (2000). These authors argue that any tool not considering the adequate geographic scale, the correct accessibility measures and the interactions between them will not capture the true relationship between space economy and public capital. We are also aware of the potential importance of spillover effects caused by transport infrastructure projects. The spatial econometric model adopted, a spatial Durbin model, allows the existence of the effects of transportation infrastructures across boundaries to be contrasted. The estimation coefficients of this model cannot be interpreted directly as additional calculations are needed to adequately incorporate these direct and indirect spillover effects.

The structure of this chapter is as follows. In section 9.2, we review the methodological issues of the production function approach, the building of the accessibility measure and the treatment of the spillover effects. In section 9.3 the empirical models and econometric matters are discussed. In section 9.4 we describe the data used and the source of the variables. In section 9.5 we present the estimation results and section 9.6 contains some conclusions and policy recommendations.

9.2 THEORETICAL BACKGROUND

In this chapter, we focus on the changes in productivity that result from increased infrastructural investments by using a primal approach. The main aim of this study is to estimate the output elasticity of road infrastructures; to achieve this objective, production function methodology is more useful than cost and profit function methodologies (Pfähler et al., 1996). We suppose there is a conventional output production function which relates real physical output (Y), to the quantity of variable inputs (X), quasi-fixed private capital input (K) and external factors as different types of public transportation infrastructures (G):

$$Y = f(X, K, G) \tag{9.1}$$

In a log-linear Cobb–Douglas specification:

$$\ln Y = \alpha_0 + \alpha_1 \ln X + \alpha_2 \ln K + \alpha_3 \ln G + \mu \tag{9.1'}$$

where μ is distributed as $[N(0, \sigma^2_{\mu} I_n)]$.

9.2.1 Accessibility

The concept of accessibility is closely related to the concepts of mobility, development, social welfare and environmental impact. Accessibility can therefore be considered as a proxy for a set of effects related or caused by transport infrastructure (Condeço-Melhorado et al., 2011). Stock capital indicators are not a satisfactory measure of the connectivity properties of the network, but only of the quantitative properties of the infrastructure. Introducing the accessibility measurement provides an advantage over these indicators as it can be understood as a measure of the potential spatial interaction and also the intensity of spillovers.

Since the main purpose of this study is to compute the impact of road infrastructures on provincial output, public transport infrastructures (*G*) were divided in two different variables: (1) public capital in the form of roads; and (2) other modes of transport. Despite the large dependence of Spanish companies on road transport,[1] other modes of transport endowments are also included to test their possible impact. However, since better information about road transport infrastructures and vehicles is available an indicator can be built that accounts for road services.

Following Fernald (1999), we suppose that road services (*RS*) depend upon the flow of services provided by the aggregate stock of government roads (*ROAD*) as well as the stock of vehicles (*VEH*) as shown in equation (9.2):

$$RS_{it} = ROAD_{it}{}^{*} VEH_{it} \tag{9.2}$$

Based on Fernald's idea we have built a measure that tries to accommodate the road stock to its degree of utilization. Following his proposal, if road infrastructure made companies more productive, sectors of the economy making heavy use of roads would benefit more from their improvements. In the road services variable, we intend to adapt this idea and apply it to provinces instead of sectors.

We have computed an alternative accessibility variable (*Acc*). The accessibility of a certain province is calculated using the population of provinces weighted by the distance travelled by road. This variable would measure the economic or market potential of a certain province, which is influenced by its own population and also by the level of activity in the surrounding regions (Gutiérrez et al., 2010). In a recent paper Holl (2012) applies a similar approach to analyse the effect of transport infrastructure investment on the productivity of Spanish manufacturing firms using company-level data.

The accessibility measure takes the form presented in equation (9.3):

$$acc_{it} = \sum_{j}^{N} \frac{POP_{jt}}{d_{ijt}} \tag{9.3}$$

where the market potential of province i (acc_{it}) depends on the volume of activity (population) of provinces j (POP_{jt}) and on the distance between provinces i and j (d_{ijt}). Instead of using Euclidian distances between mainland provinces, time-variant distances travelled by road were computed as discussed in section 9.4.

In our model we will include the accessibility measure to assess the impact of road transportation infrastructure on productivity and alternatively we will also use the stock of road transportation infrastructure and the road services variable computed previously as benchmark models. Figure 9.1 and Figure 9.2 display a map of Spain with the distribution of the three alternative variables at the beginning and the end of the period of our analysis, which starts in 1997 and ends in 2006.

9.2.2 Treatment of Spillovers

When exploring the possible spillover effects of the independent variables it should be borne in mind that different categories of public capital may not have the same spatial effects on output; that is, urban and water facilities projects may enhance economic activities in local areas, whereas communication and transport infrastructure may cause important network effects. Transport infrastructures are one of the public capital components generating the greatest interest (Cohen, 2010). Spillover effects seem especially important in this sort of project because public endowments in a region may not affect only that region, but also other geographical units connected by a transport network (Boarnet, 1998). In fact, state highways are a natural focus to test these effects since the interstate highway system in particular is designed with interstate linkages in mind (Holtz-Eakin and Schwartz, 1995).

Despite recent developments the nature of infrastructure spillovers also remains inconclusive: positive and negative spillovers have been found. To explain possible negative spillovers, we follow Boarnet (1998). Supposing an increase in public capital in region A, there would be a rise in the price of labour and capital in the region, inducing the resources to move from other regions to region A. This migration would yield a new output in region A, reducing the output in the rest of the regions. Therefore, total output in one region would depend positively on its infrastructure stock and negatively on the infrastructure stock of other regions as a result of negative output spillovers. These negative spillovers, called distributive effects by Rietveld (1994), might not arise in an analysis at a low spatial

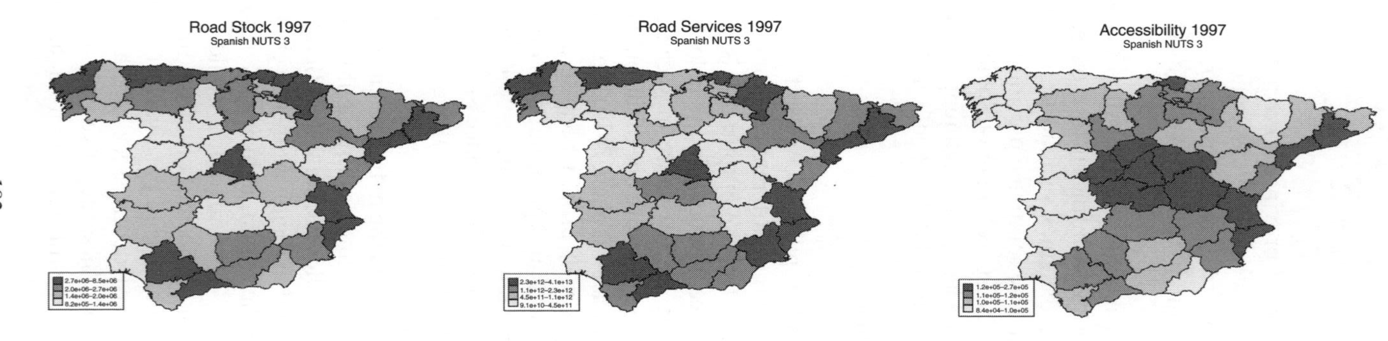

Figure 9.1 Spatial distribution of road stock, road services and accessibility measure, 1997

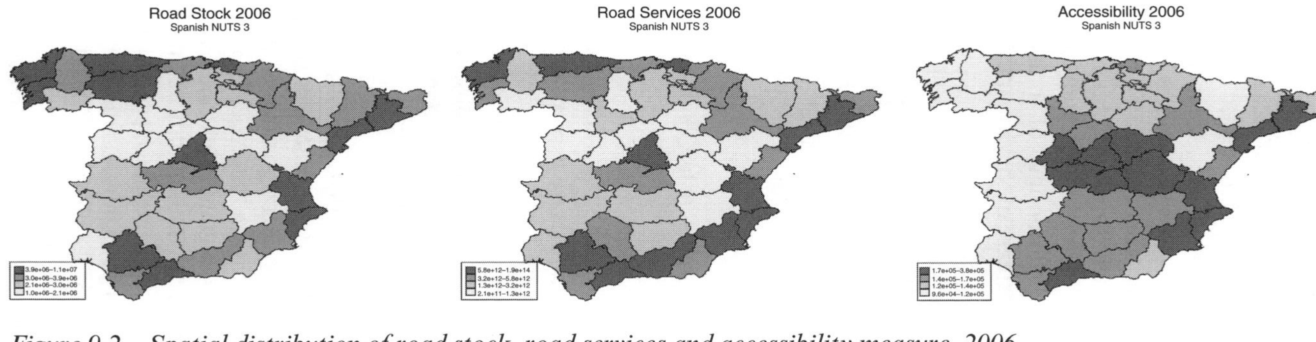

Figure 9.2 Spatial distribution of road stock, road services and accessibility measure, 2006

level. For instance, if we focused on an urban area using such low data, we might observe the building of offices or industrial facilities near a new highway; but these would have been built elsewhere if the highway did not exist leading to an overestimate of the effect.

Conversely, the foundations of the existence of positive spillovers rely on the network characteristics of transportation infrastructure in which every piece is subordinate to the entire system (Moreno and López-Bazo, 2007).[2] Road network improvements in neighbouring provinces might lead to a decrease in the transportation costs of moving inputs and final products for the economy of a particular province, which might translate into an increase in the demand for manufacturing goods and services. Congestion might also play a significant role when explaining positive spillovers; new transportation infrastructures in regions in which bottlenecks exist might improve the performance of the entire network.

In equation (9.4) the provincial Cobb–Douglas production function is augmented in two ways: by including spillover effects using the spatial lag of the independent variables, and by including the spatial lag of the dependent variable (Y):

$$\ln Y = \alpha_0 + \rho W \ln Y + \alpha_1 \ln X + \alpha_2 \ln K + \alpha_3 \ln G + \gamma_1 W \ln X + \gamma_2 W \ln K + \gamma_3 W \ln G + \varepsilon \quad (9.4)$$

where Y is the output of province, X is a matrix containing variable inputs, K contains quasi-fixed input private capital, G contains public transport infrastructure variables and ε is the error term. α, ρ and γ are the different parameters to be estimated, while W is the row standardized N-by-N spatial weight matrix with $W_{ij} > 0$ when observation j is a spatial neighbour to observation i. To test the consistency of the results, models are estimated using two different weighting matrices (W) explained in section 9.4.

The specification of equation (9.4) leads to what has been labelled a spatial Durbin model (SDM), which includes both the spatial lags of the dependent variable and those corresponding to the independent variables (LeSage and Pace, 2009). From a theoretical point of view, shocks in the production of neighbouring regions might increase the demand for products in the region of reference. In international macroeconomics when an economic boom produces an increase in the output of a country such as the United States of America, simultaneous increases in outputs in other countries are observed. Open economy models frequently have problems explaining why business cycles are so closely related among countries. According to Baxter and Farr (2005), this frequently requires implausibly high cross-country correlations of productivity shocks. By including the spatial lag of the dependent variable we attempt to contrast the existence of spatial dependence on productivity.

9.3 MODEL SPECIFICATION

The empirical model we estimate is based on the log-linear Cobb–Douglas production function. Following the previous discussion about different spatial econometric models, we estimate a spatial Durbin model:

$$y_{it} = \alpha_0 + \rho W y_{it} + \beta_1 l_{it} + \beta_2 hk_{it} + \beta_3 k_{it} + \beta_4 acc_{it} + \beta_5 trans_{it} + \theta_1 W l_{it} + \theta_2 W hk_{it} + \theta_3 W k_{it} + \theta_4 W acc_{it} + \theta_5 W trans_{it} + \varepsilon_{it} \quad (9.5)$$

where variables on both sides of the equations are in logarithms, ε is a well-behaved error term, and subscripts i and t denote provinces and time periods, respectively. Compared to equation (9.4), this equation differentiates between labour (l) and human capital (hk) and it also includes public capital (G) separated into two variables: road accessibility (acc) and other transportation modes infrastructure stock ($trans$). As commented in the previous section, road stock and road services variables are estimated as alternative measures to the accessibility variable. Finally, spatial fixed effects (μ_i) are introduced into the model to control for all time-invariant variables.

As stated above, the above equations take into account the spatial lag of the dependent variable (Wy) and the spatial lag of the explanatory variables. Two different criteria have been used to build W.[3] W_n stands for a physical contiguity matrix, where the value is one for two bordering provinces and zero for all others. W_{d150} is another binary weighting matrix with a value of one assigned to those provinces within a radius of 150 kilometres from the centroid of the province of reference and zero for provinces beyond that distance. These matrices treat physical proximity as the main driver for the presence of spillovers.

9.4 DESCRIPTION OF DATA AND VARIABLES

In this section, we discuss the data used in the estimation of the model. Spain is a decentralized country made up of two autonomous cities (Ceuta and Melilla) and 17 autonomous communities, each with its own heritage and government. These autonomous communities correspond to NUTS 2 in the European territorial unit classification and are composed of 47 mainland provinces (NUTS 3). The autonomous communities and provinces can be considered as regional economies nested within a national system. The main feature of this system is the interdependence of Spanish provinces because the evolution of each region depends on the behaviour of neighbouring regions.[4]

We use a balanced panel dataset of 47 Spanish mainland provinces covering the period from 1997 to 2006 with 470 observations. The dependent variable gross added value measured in thousands of euros (with base year 2000) is taken from the National Statistics Institute (INE). The explanatory variables – labour force, measured in thousands of workers; and human capital, measured as the share of total employment with higher-level education (secondary school, technical college and university degrees) – were taken from the INE and BBVA Foundation-Ivie, respectively. Finally, the population variable is also from the official statistics provided by the INE.

The latest series of capital stock for the Spanish economy were obtained from BBVA Foundation-Ivie (see Mas et al., 2007) where net wealth and productive capital stock data are available for both public and private capital. Productive capital stock at constant pricing is a quantity factor that takes into account loss of efficiency as assets age and is the relevant component for productivity analysis.[5] Transportation infrastructure projects, such as ports, airports and railways, have been combined into one single variable excluding road stock since this was previously included in the accessibility variable.

Since the 1970s there has been substantial development of road transportation infrastructure in Spain and especially during the 1990s the implementation of the Infrastructure and Transport Strategic Plan caused a significant boost in investment in high-capacity networks (Delgado and Álvarez, 2007).

The measurement of the time-variant distance is carried out using a simple calculation involving two variables available in the Permanent Survey of Road Transport of Goods, issued by the Spanish Ministry of Development. The mean travel distance of goods can be retrieved by dividing the total tonne-kilometres by the transported tonnes in each origin–destination pair of provinces. The distance variable, measured in this way, reports information on the routes chosen by the transport companies and the real kilometres travelled by road. It could be seen as a proxy for the travel times between two provinces without the clear disadvantage caused by the discretionary selection of points of origin and destination. It also permits the computation of trip distances within the same province.

9.5 RESULTS

9.5.1 Spatial Durbin Model (SDM) Interpretation

Before discussing the results it is worth noting that the coefficient estimates must be interpreted carefully because they depend on model specifications. The effect of the independent variables on the dependent variable in the SDM has no straightforward interpretation and direct and indirect effects must be computed. LeSage and Pace (2009) show that the partial derivates take the form of an N-by-N matrix for each k regressor and comment on their fundamental properties. For instance, the partial derivates matrix corresponding to the accessibility regressor taken in equation (9.5) would have the following form:

$$\frac{\partial y_t}{\partial acc_t} = (I_N - \rho W)^{-1}(\beta_4 I_N + \theta_4 W) \tag{9.6}$$

These authors propose scalar summary averages to facilitate the reporting of the effects associated with the regressors; thus, direct effects measure what effect changing an independent variable has on the dependent variable of a province. Direct effects, which appear in the main diagonal of the matrix shown in equation (9.6), are their own partial derivatives and are summarized using the average of these elements of the matrix. This measure includes feedback effects, that is, those effects passing through neighbouring units and back to the unit that instigated the change. The cross-partial derivatives are named indirect effects and they measure the effect of changing an independent variable in a province on the dependent variable of all the other provinces. Indirect effects appear as off-diagonal elements and are summarized as row sum averages. Finally, total effects are computed as the sum of direct and indirect effects.

9.5.2 Comments on the Results

The results obtained through the estimation process are shown in Table 9.1, which contains the point estimates of the production function with three different settings: the stock of road infrastructure and a road services variable based on Fernald's idea, with both serving as benchmark models, and an alternative model including the accessibility variable of interest. The different specifications of the models are estimated using two alternative spatial weight matrices as described above. In Table 9.2 direct, indirect and total effect computations are reported for the SDM.

The results overall are consistent with other production function

Table 9.1 Spatial Durbin model with spatial effects

	Road stock				Road services				Accessibility			
	Wn		Wd150		Wn		Wd150		Wn		Wd150	
Variable	Coef.	t-ratio	Coef.	t-ratio	Coef.	t-ratio	Coef.	t-ratio	Coef.	t-ratio	Coef.	t-ratio
L	0.347***	12.13	0.355***	12.43	0.296***	10.36	0.299***	10.42	0.364***	12.72	0.377***	13.20
HK	−0.014	−0.97	−0.013	−0.87	−0.024*	−1.69	−0.024	−1.63	−0.006	−0.47	−0.006	−0.38
K	0.078**	2.00	0.069**	1.77	0.060	1.61	0.063*	1.71	0.066*	1.69	0.045	1.14
Road Stock	0.083***	3.23	0.078***	3.04	–	–	–	–	–	–	–	–
Road Serv	–	–	–	–	0.082***	6.32	0.078***	5.92	–	–	–	–
Accessibility	–	–	–	–	–	–	–	–	−0.011	−0.45	−0.010	0.36
Trans	−0.001	−0.02	0.001	0.13	−0.003	−0.495	−0.003	−0.52	0.001	0.20	0.002	0.35
*W*L*	−0.108*	−1.80	−0.131**	−2.24	−0.176***	−3.00	0.045	0.69	−0.087	−1.44	−0.139**	−2.35
*W*HK*	0.136***	4.81	0.103***	3.82	0.083***	2.83	0.046*	1.65	0.157***	5.94	0.127***	4.96
*W*K*	0.051	0.75	0.066	0.95	0.067	1.033	0.045	0.69	0.039	0.57	0.107	1.55
*W*RoadStock*	0.026	0.66	0.054	1.27	–	–	–	–	–	–	–	–
*W*RoadServ*	–	–	–	–	0.042*	1.93	0.057***	2.64	–	–	–	–
W Accessibility*	–	–	–	–	–	–	–	–	0.070**	1.87	0.052*	1.64
*W*Trans*	−0.014	−1.45	−0.011	−1.17	−0.026***	−2.70	−0.024***	2.53	−0.015	−1.21	−0.007	−0.78
*W*Y*	0.307***	5.35	0.318***	5.73	0.210***	3.41	0.224***	3.78	0.347***	6.27	0.359***	6.74
Corrected R^2	0.949		0.947		0.955		0.953		0.947		0.944	
Log-likelihood	1131.34		1124.67		1154.321		1147.36		1124.85		1116.84	
N. Obs.	470		470		470		470		470		470	

Notes: Spatial fixed effects not shown. * Significant at 10%. ** Significant at 5%. *** Significant at 1%

Table 9.2 Direct, indirect and total effects

		Road stock				Road services				Accessibility			
		Wn		Wd150		Wn		Wd150		Wn		Wd150	
		Coeff.	t-ratio	Coeff.	t-ratio	Coeff.	t-ratio	Coeff.	t-ratio	Coeff.	t-ratio	Coeff.	t-ratio
Direct	*L*	0.348***	12.46	0.352***	11.63	0.291***	9.76	0.294***	10.41	0.366***	12.54	0.382***	11.06
	HK	−0.003	−0.24	−0.005	−0.36	−0.020	−1.45	0.051	1.50	0.007	0.50	0.009	0.53
	K	0.09**	2.33	0.078**	2.09	0.063*	1.75	0.066*	1.86	0.071*	1.88	0.062	1.31
	RoadStock	0.086***	3.57	0.085***	3.36	–	–	–	–	–	–	–	–
	Road Serv	–	–	–	–	0.085***	6.71	0.081***	6.18				
	Accessibility	–	–	–	–	–	–	–	–	−0.005	−0.21	0.008	0.258
	Trans	−0.001	−0.23	0.001	−0.07	−0.004	−0.68	−0.005	−0.84	0.000	0.00	0.002	0.32
Indirect	*L*	0.001	0.01	−0.023	−0.30	−0.140*	−1.99	−0.139**	−2.11	0.051	0.64	0.087	1.59
	HK	0.181***	4.75	0.138***	3.96	0.096**	2.66	0.051	1.51	0.224***	6.25	0.210***	8.01
	K	0.098	1.16	0.118	1.36	0.096	1.32	0.076	1.02	0.095	1.59	0.060	0.79
	RoadStock	0.071	1.39	0.106**	2.02	–	–	–	–	–	–	–	–
	Road Serv					0.071***	3.06	0.093***	3.80	–	–	–	–
	Accessibility	–	–	–	–	–	–	–	–	0.097*	1.99	0.055*	1.71
	Trans	−0.019	−1.40	−0.014	−1.12	−0.033***	−2.82	−0.030***	−2.73	−0.018	−1.27	0.002	0.23
Total	*L*	0.348***	4.09	0.329***	3.79	0.151**	2.44	0.154**	2.06	0.418***	4.69	0.470***	8.56
	HK	0.178***	4.27	0.133***	3.43	0.076*	1.91	0.030	0.80	0.231***	6.08	0.220***	9.83
	K	0.183**	2.36	0.197**	2.39	0.159**	2.44	0.141**	2.07	0.166**	1.98	0.122**	2.26
	RoadStock	0.157***	3.03	0.192***	3.45	–	–	–	–	–	–	–	–
	Road Serv	–	–	–	–	0.157	6.45	0.175***	6.93	–	–	–	–
	Accessibility	–	–	–	–	–	–	–	–	0.092*	1.75	0.063**	2.02
	Trans	−0.021	−1.47	−0.015	−1.08	−0.037***	−2.99	−0.035***	−2.03	−0.018	−1.18	0.005	0.58

Note: * Significant at 10%. ** Significant at 5%. *** Significant at 1%

studies and indicate the existence of transport infrastructure spillovers. As explained below, there are some results shared by all the models. The different specifications of the model yield similar results regarding the output point estimates of the coefficients accompanying the regressors of interest. It is worth underlining the positive and highly significant impact of the spatial lag of the dependent variable ranging from 0.21 to 0.36. In a regional production framework, this result can be interpreted as a positive correlation between the business cycle of a province and the output of its neighbours.

However, as discussed above, inferences must be made about the effect of independent variables on the productivity of a province with regard to the direct, indirect and total effects displayed in Table 9.2. According to these results the direct impacts of labour and private capital on the aggregated output of a certain province are positive and significant. Moreover, these output elasticities are quite stable in all the models and range between 0.29 and 0.38 with respect to labour and from 0.06 to 0.09 with regard to private capital.[6] These coefficients are similar to those obtained in some of the latest studies in Spain (Márquez et al., 2010).

Computation of the average direct impact of the accessibility variable indicates that the estimated coefficient is not statistically different from zero, while the impact yielded by the models when we use the road infrastructure stock and the road services variables as benchmark cases is positive and significant. Estimations of the direct impact of other types of transport infrastructures different from roads are nonetheless not significant, regardless of the empirical specification.

In the SDM the indirect effects influence the existence and size of effects across boundaries. In the benchmark models, where the road infrastructure stocks and road services variables are employed alternatively, we find some evidence of positive spatial spillovers. In the models including the accessibility variable similar results are obtained with an estimate of 0.097 and 0.055 depending on the W matrix used to define the neighbours.

According to these results, increases in the road accessibility of a province would yield an up to 9.7 per cent positive effect on the productivity of its neighbour. When the alternative measures are used, the impacts on productivity range from 7 per cent to 10.6 per cent. For the remaining modes of transportation we failed to find clear evidence of negative spillovers. Negative estimated coefficients are statistically significant in the road services model but these remarkable results are not achieved in the other two specifications. Following the interpretations presented in section 2.2, improvements in the accessibility of a province by road would cause positive spillovers to other provinces by raising the quality of the road transportation network as a whole. Conversely, increased investment into ports, airports and railway infrastructure projects in a province would produce negative or zero spillovers to other regions.

The empirical approach used in this chapter provides evidence of spatial spillover effects for the different types of transportation infrastructure projects, consistent with most of the literature using Spanish provincial data. For instance, Delgado and Álvarez (2007), putting into practice a stochastic frontier approach, found positive and negative spillovers depending on the sector of the economy under review and the definition of the weighting matrix. Using a production function, Moreno and López-Bazo (2007) found the existence of negative spatial spillovers of transport infrastructure. In contrast, Álvarez et al. (2006) replicated the models used by Holtz-Eakin and Schwartz (1995) and Mas et al. (1994) using Spanish provincial data and did not find either positive or negative spillovers.

Finally, we obtain the total provincial productivity impacts of the variables by adding direct and indirect effects. As explained above, the average total effect of road infrastructure is positive, statistically significant and very similar in the benchmark settings. Nonetheless, the total effect estimated for the accessibility variable ranges from 0.063 to 0.092.

9.6 CONCLUSIONS

In this chapter we have attempted to find an alternative aggregated production function to measure the effects of road infrastructure public investments on the economy of Spain. The main contribution of the work is twofold. Firstly, the creation of an accessibility variable for a specific province, that combines the population of the province of origin with the population of other provinces weighted by road distance. Using this variable, we try to avoid the shortcomings caused by the use of transport infrastructure stock variables. To contrast the validity of this approach we have also included two alternative measures as benchmarks for the model. Secondly, the empirical models include spatial lags of the independent variables and also of the dependent variable, which is not a common approach in regional productivity analysis literature. Controlling for spatial dependence, in the form of public capital spillovers, we find strong evidence of the positive impact of better road accessibility on the private economy of a province.

The evidence obtained indicates that the spillover effects of road infrastructure and accessibility are larger than its direct impacts. As a consequence, improving accessibility in one spatial unit would increase the productivity of neighbouring areas even more than in the spatial unit where the infrastructure is located. A policy implication from this model seems to support the idea that road transportation infrastructure should be planned by the national government. This conclusion could be of major importance in the Spanish political context because both regional and provincial governments participate with the national government in the decisions on where to raise the infrastructure endowments. From a methodological point of view, this result provides new findings on the importance of including spillover effects on regional production functions in line with previous literature.

NOTES

1. According to the National Statistics Institute, road transport was chosen in more than 77 per cent of freight movements in Spain in 2007.
2. We are aware of Braess's paradox, which states that an increase in the capacity of a transportation network might reduce its overall performance. However, we assume that this phenomenon is less likely to be felt on an aggregate level than the positive effects of network improvement.
3. The weighting matrices have been row normalized following standard practice in the spatial econometrics literature. After this transformation, the sum of all elements in each row equals one. Note that the row elements of a spatial weighting matrix show the effect on a particular unit of all other units.
4. Márquez and Hewings (2003) analyse regional competition between Spanish regions (NUTS 2).
5. The computations of the productivity of capital stock are obtained using a new methodology applied to Spanish capital stock estimates that is based on two OECD manuals (Schreyer, 2001; Schreyer et al., 2003).
6. Elhorst (2010) emphasizes that empirical studies usually find significant differences among the coefficient estimates from models with and without spatial fixed effects. Models that include spatial fixed effects use time-series variations of the data, whereas models that do not control for spatial fixed effects utilize cross-sectional components of the data. Models of the first type tend to give short-term estimates, and models without controls for spatial fixed effects tend to give long-term estimates (Baltagi, 2005).

REFERENCES

Álvarez, A., C. Arias and L. Orea (2006), 'Econometric testing of spatial productivity spillovers from public capital', *Hacienda Pública Española*, **178**, 9–21.

Aschauer, D.A. (1989), 'Is public expenditure productive?', *Journal of Monetary Economics*, **23**, 177–200.

Baltagi, B.H. (2005), *Econometric Analysis of Panel Data*, 4th edn, Chichester: Wiley.

Baxter, M. and D.D. Farr (2005), 'Variable capital utilization and international business cycles', *Journal of International Economics*, **65**, 335–347.

Boarnet, M.G. (1998), 'Spillovers and the locational effects of public infrastructure', *Journal of Regional Science*, **38**, 381–400.

Cohen, Jeffrey P. (2010), 'The broader effects of transportation infrastructure: spatial econometrics and productivity approaches', *Transportation Research Part E*, **46**, 317–326.

Cohen, J.P. and C.J. Morrison Paul (2004), 'Public infrastructure investment: interstate spatial spillovers and manufacturing costs', *Review of Economics and Statistics*, **86**, 551–560.

Condeço-Melhorado, A., J. Gutierrez and J.C. Garcia-Palomares (2011), 'Spatial impacts of road pricing: accessibility, regional spillovers and territorial cohesion', *Transportation Research Part A*, **45**, 185–203.

Delgado, M.J. and I. Álvarez (2007), 'Network infrastructure spillover in private productive sectors: evidence from Spanish high capacity roads', *Applied Economics*, **39**, 1583–1597.

Elhorst, J.P. (2010), 'Applied spatial econometrics: raising the bar', *Spatial Economic Analysis*, **5** (1), 9–28.

Fernald, J. (1999), 'Roads to prosperity? Assessing the link between public capital and productivity', *American Economic Review*, **89** (3), 619–638.

Forkenbrock, D. and N. Foster (1990), 'Economic benefits of a corridor highway investment', *Transportation Research Part A: General*, **24** (4), 303–312.

Forslund, U. and B. Johansson (1995), 'Assessing road investments: accessibility changes, cost benefits and production effects', *Annals of Regional Science*, **29**, 155–174.

Gutiérrez, J., A. Condeço-Melhorado and J.C. Martín (2010), 'Using accessibility indicators and GIS to assess spatial spillovers of transport infrastructure investment', *Journal of Transport Geography*, **18** (1), 141–152.

Holl, A. (2012), 'Market potential and firm-level productivity in Spain', *Journal of Economic Geography*, **12** (6), 1191–1215.

Holtz-Eakin, D. and A. Schwartz (1995), 'Spatial productivity spillovers from public infrastructure: evidence from state highways', *International Tax and Public Finance*, **2**, 459–468.

Hulten, C. (2004), 'Transportation infrastructure: productivity and infrastructures', paper prepared for the 132nd Round Table of the European Conference of Ministers of Transport at the Joint OECD/EMCT Transport Research Center, Paris, 2–3 December.

LeSage J. and R.K. Pace (2009), *Introduction to Spatial Econometrics*, Boca Raton, FL: CRC Press.

Márquez, M.A. and G.J.D. Hewings (2003), 'Geographical competition between regional economies: the case of Spain', *Annals of Regional Science*, **37** (4), 559–580.

Márquez, M.A., J. Ramajo and G. Hewings (2010), 'A spatio-temporal econometric model of regional growth in Spain', *Journal of Geographical Systems*, **12** (2), 207–226.

Mas, M., J. Maudos, F. Pérez and E. Uriel (1994), 'Capital público y productividad en las regiones españolas', *Moneda y crédito*, **198**, 163–206.

Mas, M., F. Pérez and E. Uriel (2007), 'El stock y los servicios del capital en España y su distribución territorial (1964–2005)', Nueva metodología, Bilbao: Fundación BBVA.

Mera, K. (1973), 'Regional production functions and social overhead capital: an analysis of the Japanese case', *Regional Science and Urban Economics*, **3** (2), 157–185.

Mikelbank, B. and R. Jackson (2000), 'The role of space in public capital research', *International Regional Science Review*, **23**, 235–258.

Moreno, R. and E. López-Bazo (2007), 'Returns to local and transport infrastructure under regional spillovers', *International Regional Science Review*, **30** (1), 47–71.
Munnell, A.H. and L.M. Cook (1990), 'How does public infrastructure affect regional economic performance?', *New England Economic Review*, September, 11–33.
Oosterhaven, J. and T. Knaap (2003), 'Spatial economic impacts of transport infrastructure investments', in A. Pearman, P. Mackie and J. Nellthorp (eds), *Transport Projects. Programmes and Policies: Evaluation Needs and Capabilities*, Aldershot: Ashgate, pp. 87–105.
Pfähler, W., U. Hofmann and W. Bönte (1996), 'Does extra public infrastructure capital matter? An appraisal of empirical literature', *Public Finance Analysis*, New Series, **Bd 53**, H.1 (1996/1997), 68–112.
Rietveld, P. (1994), 'Spatial economic impacts of transport infrastructure supply', *Transportation Research Part A*, **28** (4), 329–341.
Rietveld, P. and P. Nijkamp (2000), 'Transport infrastructure and regional development', in J.B. Polak and A. Heertje (eds), *Analytical Transport Economics. An International Perspective*, Cheltenham, UK and Northampton, MA, USA: Edward Elgar Publishing, pp. 208–232.
Schreyer, P. (2001), 'The OECD productivity manual: a guide to the measurement of industry-level and aggregate productivity', *International Productivity Monitor*, **2**, 37–51.
Schreyer, P., P.E. Bignon and J. Dupont (2003), 'OECD capital services estimates: methodology and a first set of results', OECD Statistics Working Papers 2003/6, OECD Publishing.

10. Location, accessibility and firm-level productivity in Spain

Adelheid Holl

10.1 INTRODUCTION

For economists and policy-makers, private sector productivity is important as it is closely related to economic growth and international competitiveness. The productivity literature has so far documented large productivity differences across firms even within narrowly defined industries (see e.g. Syverson, 2011) but also with significant disparities across locations. Specifically, there has been a recent surge in interest in the effect of urban size and density on firm-level productivity. This literature generally agrees that agglomeration has a positive effect on productivity (Rosenthal and Strange, 2004; Melo et al., 2009; Puga, 2010). For example, Ciccone and Hall (1996) show for the United States that higher employment density increases labour productivity. Ciccone (2002) and Brülhart and Mathy (2008) provide evidence for European regions that productivity is higher in regions with higher employment density. Combes et al. (2010) show that local employment density enhances productivity among French firms. Martin et al. (2011) find for French firms a positive effect of same industry spatial concentration. Also using French data, Combes et al. (2012) show how plants in large cities are more productive than plants elsewhere. Building on the seminal work of Marshall (1920), it is argued that in larger cities firms have more other firms nearby and that this generates positive externalities in terms of knowledge spillovers, labour market pooling and input sharing, and consequently improves firms' productivity.

The benefits of agglomeration are intrinsically linked to transport infrastructure (Eberts and McMillen, 1999; Holl, 2004; Venables, 2007; Graham 2007a, 2007b). Better transport infrastructure reduces spatial frictions and can also extend the spatial scope over which agglomeration benefits are obtained. Few empirical studies to date, however, have directly tested for transport infrastructure-based accessibility measures of agglomeration in firm-level productivity studies. Lall et al. (2004) find that road access to markets is an important determinant of plant-level productivity among Indian manufacturing firms. Graham (2007b) estimates translog firm-level production functions for UK firms and finds positive effects of road distance-based market potential for most services industries as well as for manufacturing. Holl (2012) provides empirical evidence for Spain on a positive impact of road infrastructure-based market potential on firm-level productivity. Gibbons et al. (2012) find positive effects of accessibility to employment along the road network on labour productivity for UK firms.

In this chapter the effect of different location characteristics on firm-level productivity is compared including local population size and local population density, as well as transport accessibility measured via market potential. The use of accessibility measures permits the direct focus on the interactions between economic agents across space facilitated via transportation systems and has the advantage that it does not limit the source of location benefits to administrative boundaries.

To measure the impact on firm-level productivity, first, total factor productivity (TFP) is calculated using data from the SABI (Sistema de Análisis de Balances Ibéricos) database generated by INFORMA and Bureau Van Dyck, which provides detailed location information for a large panel of firms for 1997–2005. I calculate TFP using the Levinsohn and Petrin (2003) approach to control for endogeneity problems in TFP arising from unobserved productivity shocks. Next, the relationship between the three location characteristics and firm-level productivity is examined. In estimating the effects of location characteristics on firm-level productivity it is, however, also important to take into account their potential endogeneity through the mobility of workers and the sorting and selection of firms via their location decision. In order to estimate a causal effect, different specifications including fixed effects as well as instrumental variable estimations are used. The results indicate a statistically significant positive effect for location characteristics and especially for transport infrastructure-based market potential on firm-level productivity. The results further show that at a geographically detailed level of analysis, accessibility-based measures better capture the benefits of location than measures based on administrative boundaries.

The chapter is organized as follows. Section 10.2 covers data and model specification. Section 10.3 presents the results. Section 10.4 concludes.

10.2 DATA, ESTIMATION EQUATION AND VARIABLE DEFINITION

10.2.1 Data

The dataset used to calculate firm-level productivity is the SABI (Sistema de Análisis de Balances Ibéricos) database. SABI is generated by INFORMA and Bureau Van Dyck and contains financial accounts for over 1.2 million Spanish companies. The analysis is restricted to manufacturing firms. In SABI, slightly more than 165 000 firms belong to the manufacturing sector. The database contains exhaustive balance sheet information as well as other firm characteristics and detailed location information (geographic coordinates). The database contains information on nearly the complete population of firms, excluding only very small firms. Multi-plant companies have been dropped because the information available from the SABI database does not permit outputs and inputs to be assigned correctly among plants of multi-plant companies. In addition, firms that have relocated over the period of analysis and firms that have changed their industrial sector have been excluded. To identify those firms, I have extracted information on firms' previous location and sector from all earlier editions of the SABI database. In this way, I have identified 10 583 firms that changed location beyond the city level over the period of analysis and 12 795 firms that changed their industrial sector.

Finally, in cleaning the dataset, observations with missing, negative or null values for value added, employment or stock of capital, as well as with missing information for other independent variables have been dropped. This leaves me with 315 560 observations in the final sample consisting of an unbalanced panel of 70 950 mainland manufacturing firms.[1] Table 10A.1 in the Appendix provides information on the distribution of firms by year in the final sample used for estimation.

To calculate the transport-based accessibility measure, the Spanish road network has been entered into a geographical information system (GIS) annually for the period of analysis. The historical evolution is based on detailed information obtained from the Ministry of Public Works regarding the opening of new motorway segments. This information has been combined with information from the annual official roadmaps published by the Ministry of Public Works. For each new motorway road segment in the GIS database, the associated tabular data contains information on the year it was opened to traffic (see also Holl, 2007).

10.2.2 Model Specification

A reduced-form equation is estimated where firm-level total factor productivity (TFP) of firm *i* in sector *s*, province *p* and year *t* is a function of firm-specific characteristics c_{it}, the municipality characteristics m_{jt} where the company is located, sector fixed effects s_s, province fixed effects p_p, and year fixed effects γ_t:

$$\log TFP_{ispt} = \alpha + \beta_1 c_{it} + \beta_2 m_{jt} + s_s + p_p + \gamma_t + e_{ispt} \quad (10.1)$$

To calculate TFP, the standard approach of Levinsohn and Petrin (2003) is applied. This approach uses intermediate inputs as a proxy for unobserved productivity shocks to account for the possible endogeneity arising from such unobserved shocks. The Stata routine levpet provided by Petrin et al. (2004) is used for the estimation of the production function. Results from the estimation of the production function are not included here, but are available upon request.

Firm-level control variables include foreign ownership and the age of the company and controls for firm size based on total deflated assets.[2] Estimations further include industry fixed effects to control for the different sector distribution of firms across locations and for different industry impacts on the performance of firms.[3] Year fixed effects and province fixed effects are included to control for year and region-specific effects. Provinces relate fairly closely to labour market areas in Spain and regions with relatively better-developed human capital are likely to also have higher productivity firms. Including province fixed effects thus accounts for any such unobserved time-invariant location attributes common to the labour market areas that may affect firm-level total factor productivity. The province fixed effects help to identify the effects of location characteristics separately from effects of unobserved spatial heterogeneity common to the province level.

The main variables of interest are the characteristics of the municipality where the company is located. I first test for the effect of local population size and local population density defined as the ratio of total municipality population to the total municipality land area.

Furthermore, a transport accessibility-based market potential (*mp*) measure is calculated as:

$$mp_{jt} = pop/radius_{jt} + \sum_{k \in L_{573}} \frac{pop_{kt}}{d_{jkt}} \quad (10.2)$$

In line with the 'effective density' measure in Graham (2007a, 2007b) and Graham and Kim (2008), this is the distance-discounted sum of population in all other municipalities in the destination set L573, defined as the 573 largest Spanish cities with more than 10 000 inhabitants, plus a proxy for own municipality density measured as municipality population divided by the average municipality radius assuming a circular municipality area. d_{jk} is the distance between municipality j and k and is based on shortest path travel times along the real road network and measured in minutes.[4] The measure covers over 75 per cent of the total Spanish peninsular population.

The three location characteristics are potentially endogenous to firm-level productivity. On the one hand, a potential endogeneity source could arise from firm and worker mobility. Higher-productivity firms may self-select into larger cities, denser markets and areas with higher market potential through their initial location decision (Baldwin and Okubo, 2006; Nocke, 2006) and consequently attract more workers to that location. On the other hand, the placement of road investment may not be independent of local productivity levels.

This potential endogeneity problem is addressed using an instrumental variables approach. As in Holl (2012), I follow the recent related literature which has used historical variables and variables related to geology as sources of exogenous cross-section variation for current location characteristics. Historical population values have been used arguing that the factors that played a role in the long past are uncorrelated to the factors affecting firms' current productivity shocks (Ciccone, 2002; Combes et al., 2010). Here the municipality population data for 1900 provided by the BBVA Foundation and the Valencian Institute of Economic Research (Azagra and Chorén, 2006) is used. This data is also used to calculate a market potential measure for 1900 as the sum of own population plus the population of other areas weighted by the inverse of the geodesic distance. Local terrain ruggedness has determined settlement patterns in the past and is thus related to current population patterns but is no longer a factor influencing modern manufacturing productivity (Combes et al., 2010). As in Holl (2012), I use the national digital terrain model (MDT200) with a 200 metre elevation grid provided by the National Geographic Institute and calculate the Riley et al. (1999) terrain ruggedness index. This index gives a summary statistic of differences in metres in elevation and captures small-scale topographic heterogeneity. Finally, also as in Holl (2012), I use the distance to the 1760 postal route network as further instrument. Settlement patterns over the past have also been determined by such historic transport routes.

10.3 RESULTS

Estimation results are presented in Tables 10.1 and 10.2. Results in Table 10.1 are based on OLS estimation and should be seen as benchmark estimates because the estimates could be biased if the location variables are indeed endogenous. Firm-level TFP is regressed separately on the three location characteristics together with firm-level controls, industry fixed effects, province fixed effects and year fixed effects.

Local population size shows a significant positive effect (column 1). The elasticity of productivity with respect to population size is, however, small with a doubling of local population size being associated with only a 0.7 per cent increase in firm-level total factor productivity. The elasticity of productivity with respect to population density is slightly higher, associated with approximately a 1.2 per cent increase in firm-level TFP (column 3). These results are unchanged when industry–year fixed effects and province–year fixed effects are included as in columns (2) and (4) respectively. The industry–year fixed effects and province–year fixed effects control any shocks common either to sector or provinces stemming from unobserved factors that could account for differences in TFP.

Market potential shows a markedly higher coefficient (columns 5 and 6). A doubling of market potential is associated with almost 8 per cent higher productivity. In columns (7) and (8) local population density and market potential that excludes own municipality population density are included. Both variables show a positive and significant relation to firm-level TFP productivity, but again the coefficient of market potential is markedly higher than the coefficient for local population density. To put the results into perspective, Rosenthal and Strange (2004) in their survey of the relevant literature conclude that productivity increases by 3 to 8 per cent when regional population density doubles. Combes et al. (2011) place the range somewhat lower at 2 to 5 per cent. Martin et al. (2011) show that the effect on productivity depends on the spatial aggregation of the location variable. For France they find an elasticity of 1.96 per cent for localization economies based on the smaller employment areas and an elasticity of 4.17 per cent for localization economies based on *départements* which are larger on average. The results are also consistent with Melo and Graham (2009) who find for UK individual wages that distance-based market potential has a larger impact than local employment density.

As for firm-level controls, foreign ownership is associated with higher TFP as is age of the company. Furthermore, firm size is also positively associated with higher TFP.[5] Table 10.2 shows the results from instrumental variable estimations to control for the potential endogeneity of the three location characteristics. Columns (1) and (2) are based again

Table 10.1 OLS estimates of the effect of local population, local population density and market potential on firm-level TFP

	(1)	(2)	(3)	(4)	(5)	(6)	(7)	(8)
log (population)	0.007***	0.007***						
	(0.002)	(0.002)						
log (population density)			0.012***	0.012***			0.009***	0.008***
			(0.002)	(0.002)			(0.002)	(0.002)
log (total market potential)					0.079***	0.079***		
					(0.016)	(0.016)		
log (market potential without local population density)							0.064***	0.064***
							(0.013)	(0.013)
Firm-level controls:								
Foreign ownership dummy	0.172***	0.172***	0.172***	0.171***	0.172***	0.171***	0.171***	0.171***
	(0.017)	(0.017)	(0.017)	(0.017)	(0.017)	(0.017)	(0.017)	(0.017)
log (age)	0.057***	0.056***	0.056***	0.056***	0.057***	0.056***	0.056***	0.056***
	(0.002)	(0.002)	(0.002)	(0.002)	(0.002)	(0.002)	(0.002)	(0.002)
Size 2nd quintile	0.154***	0.154***	0.154***	0.154***	0.154***	0.154***	0.153***	0.154***
	(0.005)	(0.005)	(0.005)	(0.005)	(0.005)	(0.005)	(0.005)	(0.005)
Size 3rd quintile	0.279***	0.280***	0.279***	0.280***	0.279***	0.280***	0.278***	0.279***
	(0.006)	(0.006)	(0.006)	(0.006)	(0.006)	(0.006)	(0.006)	(0.006)
Size 4th quintile	0.416***	0.418***	0.417***	0.418***	0.416***	0.417***	0.416***	0.417***
	(0.006)	(0.007)	(0.006)	(0.007)	(0.007)	(0.007)	(0.007)	(0.007)

Table 10.1 (continued)

	(1)	(2)	(3)	(4)	(5)	(6)	(7)	(8)
Size 5th quintile	0.710***	0.711***	0.711***	0.712***	0.709***	0.710***	0.709***	0.710***
	(0.011)	(0.011)	(0.011)	(0.011)	(0.011)	(0.011)	(0.011)	(0.011)
Time fixed effects	Y	N	Y	N	Y	N	Y	N
Industry fixed effects	Y	N	Y	N	Y	N	Y	N
Province fixed effects	Y	N	Y	N	Y	N	Y	N
Industry × year fixed effects	N	Y	N	Y	N	Y	N	Y
Province × year fixed effects	N	Y	N	Y	N	Y	N	Y
Observations	315 560	315 560	315 560	315 560	315 560	315 560	315 560	315 560
R-squared	0.487	0.490	0.487	0.490	0.487	0.491	0.487	0.491

Note: Robust standard errors corrected for clustering at the municipality level are reported in parentheses. Significant coefficients are indicated by ***, **, *, for significance at the 1%, 5% and 10% level, respectively.

Table 10.2 IV estimates of the effect of local population, local population density and market potential on firm-level TFP

	(1)	(2)	(3)	(4)	(5)	(6)	(7)	(8)
log (population)	0.004**	0.004**						
	(0.002)	(0.002)						
log (population density)			0.008**	0.008**			0.011*	0.011*
			(0.003)	(0.003)			(0.006)	(0.006)
log (total market potential)					0.067***	0.067***		
					(0.026)	(0.026)		
log (market potential without local population density)							0.093***	0.091***
							(0.032)	(0.032)
Firm-level controls:								
Foreign ownership dummy	0.172***	0.172***	0.172***	0.171***	0.172***	0.171***	0.171***	0.171***
	(0.017)	(0.017)	(0.017)	(0.017)	(0.017)	(0.017)	(0.017)	(0.017)
log (age)	0.057***	0.057***	0.057***	0.056***	0.057***	0.056***	0.056***	0.056***
	(0.002)	(0.002)	(0.002)	(0.002)	(0.002)	(0.002)	(0.002)	(0.002)
Size 2nd quintile	0.153***	0.154***	0.154***	0.154***	0.154***	0.154***	0.154***	0.154***
	(0.005)	(0.005)	(0.005)	(0.005)	(0.005)	(0.005)	(0.005)	(0.005)
Size 3rd quintile	0.279***	0.279***	0.279***	0.280***	0.278***	0.279***	0.279***	0.280***
	(0.006)	(0.006)	(0.006)	(0.006)	(0.006)	(0.006)	(0.006)	(0.006)
Size 4th quintile	0.416***	0.417***	0.416***	0.417***	0.416***	0.417***	0.416***	0.417***
	(0.006)	(0.006)	(0.006)	(0.006)	(0.006)	(0.006)	(0.006)	(0.007)
Size 5th quintile	0.709***	0.710***	0.710***	0.711***	0.709***	0.710***	0.709***	0.710***
	(0.010)	(0.010)	(0.010)	(0.010)	(0.011)	(0.011)	(0.011)	(0.011)
Time fixed effects	Y	N	Y	N	Y	N	Y	N
Industry fixed effects	Y	N	Y	N	Y	N	Y	N

Table 10.2 (continued)

	(1)	(2)	(3)	(4)	(5)	(6)	(7)	(8)
Province fixed effects	Y	N	Y	N	Y	N	Y	N
Industry × year fixed effects	N	Y	N	Y	N	Y	N	Y
Province × year fixed effects	N	Y	N	Y	N	Y	N	Y
First stage F-test	110.8	110.6	177.5	177.1	94.4	94.4	34.1/13.3	34.7/13.5
First stage R^2	0.669	0.621	0.631	0.578	0.865	0.848	0.581/0.820	0.520/0.805
Stock–Wright S statistic (p-value)	0.008	0.008	0.004	0.004	0.000	0.000	0.000	0.000
P-value of Hansen J	0.130	0.139	0.326	0.337	0.160	0.172	0.163	0.171
P-value of endogeneity test	0.139	0.145	0.434	0.424	0.919	0.973	0.544	0.575
Observations	315 560	315 560	315 560	315 560	315 560	315 560	315 560	315 560
R-squared	0.487	0.252	0.487	0.252	0.487	0.253	0.487	0.253

Notes:
Robust standard errors corrected for clustering at the municipality level are reported in parentheses. Significant coefficients are indicated by ***, **, *, for significance at the 1%, 5% and 10% level, respectively. Instruments in column (1)–(2): terrain ruggedness index, 1900 market potential, land area; (3)–(4): terrain ruggedness index, 1900 market potential; (5)–(6) distance to 1760 postal routes, terrain ruggedness index; 1900 market potential; (7)–(8): distance to 1760 postal routes, 1900 population, 1900 market potential.
Craig–Donald F-statistic to test for weak instruments exceeds the Stock and Yogo's weak ID tests critical values at 5% maximal IV relative bias (see Stock and Yogo, 2002) in all estimations. In columns (2), (4), (6), and (8) industry–year fixed effects have been partialled out.

on local population size, with industry, province and year fixed effects in column (1) and industry–year fixed effects and province–year fixed effects in column (2). Accounting for endogeneity of local population leads to an even slightly smaller coefficient of about 0.004, indicating that a doubling of local population size is associated with 0.4 per cent higher firm-level TFP.

Columns (3) and (4) present the results for local population density. As with population size, the coefficients in the IV estimations are somewhat lower. The same applies to total market potential (columns 5 and 6). However, even when instrumented, market potential still shows a markedly higher coefficient compared to local population size and density. The IV estimates indicate a TFP increase of about 6.7 per cent when market potential is doubled. Including local population density together with market potential that excludes the own municipality population density proxy in columns (7) and (8) shows that the former raises productivity by about 1.1 per cent and the latter by slightly over 9 per cent.

The overall lower estimates for municipality population size and municipality population density may be related to the fact that municipalities are very small geographical units. Location at this fine-grained spatial level may only capture part of the productivity benefits of economic agglomeration. In contrast, market potential has the advantage of taking into account spatial spillovers beyond the boundaries of the spatial unit. The results provide support for the argument that the level of geographical aggregation is relevant when using administrative spatial units. However, the differences in the size of the estimates could also be influenced by some form of the modifiable areal unit problem (MAUP) (Openshaw and Taylor, 1979).[6]

To assess the validity of instruments in the 2SLS estimations, a number of test statistics are presented in Table 10.2. The F-tests for joint significance of the instruments are high in all cases and well above the typical threshold. Also the first stage R^2s show that instruments provide a good fit in the first stage, particularly in the case of the estimations with market potential. The statistically significant values of the Stock–Wright (2000) S statistic, which is robust to the presence of weak instruments (see Baum et al., 2007), further indicate that the instruments are valid. Since estimations include more instruments than endogenous variables, the Hanson J test for over-identifying restrictions can further be used to indicate whether the instruments are exogenous assuming that at least one of the instruments is valid. In all specifications the hypotheses that the instruments are valid is not rejected.

Results from an endogeneity test are reported for all specifications. This tests the hypothesis that the location characteristics used in the models are exogenous. The Durbin–Wu–Hausman test statistics are not significant in any of the specifications. This indicates that exogeneity can not be rejected in these models.

Finally, model spatial autocorrelation was assessed calculating Moran's I (Moran, 1950) of municipality average firm-level residuals. The Moran I statistics range between 0.015 and 0.023 and indicate some weak residual spatial correlation for models (1) to (6). However, the residuals of models (7) and (8), both in the OLS case as with IV, are not significantly spatially correlated across municipalities. This suggests that these models incorporate the most relevant factors accounting for spatial patterns in total factor productivity differences.[7]

10.4 CONCLUSION

This chapter has analysed the relation between location characteristics and firm-level total factor productivity. The results show that firm-level productivity is significantly related to the characteristics of the local environment. Specifically, there is a significant positive relation between transport accessibility-based market potential and firm-level total factor productivity.

Overall, accessibility-based market potential shows a stronger relation with firm-level productivity than municipality population size or municipality population density. Its coefficient is in the upper range of results recently reported in the literature on agglomeration economies and close to the estimates in Holl (2012) based on more aggregated location information. The results indicate that the advantages of location are not limited to the administrative boundaries of municipalities and that transport networks extend productivity benefits beyond localities.

In this chapter, agglomeration and accessibility measures are based on municipal population data. In further research this could be replaced with information on the co-location of firms or with local employment data both in the same sector and in other sectors in order to analyse effects depending on the composition of the local productive structure. Furthermore, testing for accessibility measures with different distance decays or different distance cut-offs may provide additional indications of the spatial extent of these benefits (Melo and Graham, 2009; Drucker, 2012).

ACKNOWLEDGEMENTS

Financial support from the Spanish Ministerio de Ciencia e Innovación [ECO2010–17485] and CSIC [200910I105] is gratefully acknowledged.

NOTES

1. On average, I have 4.4 years of observations per firm. Firms in the islands and Ceuta and Melilla are not included
2. Unfortunately, the SABI database does not include information on the workforce composition, which would allow correcting of the TFP measure for labour skill composition effects.
3. Sector information is available at the four-digit industry level. Estimations presented here are based on a grouping into 20 manufacturing sectors. Alternatively, I have also used four-digit industry dummies which lead to qualitatively similar results.
4. Here a distance decay parameter of one is implicitly assumed.
5. Note that, without controlling for firm size, neither population size nor population density are significant. Since firm size is uneven across different locations, omitting controlling for firm size may bias the results for the coefficients of the location variables.
6. See Briant et al. (2010) for an evaluation of the magnitude of distortion arising from the choice of a specific zoning system.
7. Note that total factor productivity itself is significantly spatially correlated (Moran I = 0.196, significant at the 1 per cent level). See Fischer et al. (2010), for example, for a careful analysis of spatial variation in labour productivity in Europe adopting a spatial regression approach.

REFERENCES

Azagra Ros, J. and P. Chorén Rodríguez (2006), 'La localización de la población española sobre el territorio: un siglo de cambios', Un estudio basado en series homogéneas (1900–2001), Fundación BBVA.

Baldwin, R.E. and T. Okubo (2006), 'Heterogeneous firms, agglomeration and economic geography: spatial selection and spatial sorting', *Journal of Economic Geography*, **6** (3), 323–346.

Baum, C.F., M.E. Schaffer and S. Stillman (2007), 'Enhanced routines for instrumental variables/generalized method of moments estimation and testing', *Stata Journal*, **7** (4), 465–506.

Briant, A., P. Combes and M. Lafourcade (2010), 'Dots and boxes: do the size and shape of spatial units jeopardize economic geography estimations?', *Journal of Urban Economics*, **67** (3), 287–302.

Brülhart, M. and N.A. Mathy (2008), 'Sectoral agglomeration economies in a panel of European regions', *Regional Science and Urban Economics*, **38**, 348–362.

Ciccone, A. (2002), 'Agglomeration effects in Europe', *European Economic Review*, **46** (2), 213–227.

Ciccone, A. and R.E. Hall (1996), 'Productivity and the density of economic activity', *American Economic Review*, **86**, 54–70.

Combes, P.P., G. Duranton and L. Gobillon (2011), 'The identification of agglomeration economies', *Journal of Economic Geography*, **11** (2), 253–266.
Combes, P.P., G. Duranton, L. Gobillon, D. Puga and S. Roux (2012), 'The productivity advantages of large cities: distinguishing agglomeration from firm selection', *Econometrica*, **80** (6), 2543–2594.
Combes, P.P., G. Duranton, L. Gobillon and S. Roux (2010), 'Estimating agglomeration effects with history, geology, and worker fixed effects', in Edward L. Glaeser (ed.), *Agglomeration Economics*, Chicago, IL: University of Chicago Press / National Bureau of Economic Research, pp. 15–66.
Drucker, J. (2012), 'The spatial extent of agglomeration economies: evidence from three US manufacturing industries', Working Paper CES 12–01, Center for Economic Studies, United States Census Bureau.
Eberts, R.W. and D.P. McMillen (1999), 'Agglomeration economies and urban public infrastructure', in P. Cheshire and E.S. Mills (eds), *Handbook of Urban and Regional Economics*, Vol. 3, New York: North-Holland, pp. 1455–1495.
Fischer, M.M., M. Bartkowska, A. Riedl, S. Sardadvar and A. Kunnert (2010), 'The impact of human capital on regional labor productivity', in M. Fischer and A. Getis (eds), *Handbook of Applied Spatial Analysis*, Berlin: Springer, pp. 585–597.
Gibbons, S., T. Lyytikäinen, H. Overman and R. Sanchis-Guarner (2012), 'New road infrastructure: the effects on firms', SERC Discussion Paper 117.
Graham, D.J. (2007a), 'Agglomeration, productivity and transport investment', *Journal of Transport Economics and Policy*, **41** (3), 317–343.
Graham, D.J. (2007b), 'Variable returns to agglomeration and the effect of road traffic congestion', *Journal of Urban Economics*, **62**, 103–120.
Graham, D.J. and H.Y. Kim (2008), 'An empirical analytical framework for agglomeration economies', *Annals of Regional Science*, **42**, 267–289.
Holl, A. (2004), 'Transport infrastructure, agglomeration economies, and firm birth: empirical evidence from Portugal', *Journal of Regional Science*, **44** (4), 693–712.
Holl, A. (2007), 'Twenty years of accessibility improvements: the case of the Spanish motorway building programme', *Journal of Transport Geography*, **15** (4), 286–297.
Holl, A. (2012), 'Market potential and firm-level productivity in Spain', *Journal of Economic Geography*, **12** (6), 1191–1215.
Lall, S., Z. Shalizi and U. Deichmann (2004), 'Agglomeration economies and productivity in Indian industry', *Journal of Development Economics*, **73**, 643–673.
Levinsohn, J. and A. Petrin (2003), 'Estimating production functions using inputs to control for unobservables', *Review of Economic Studies*, **70** (2), 317–341.
Marshall, A. (1920), *Principles of Economics*, 8th edn, London: Macmillan.
Martin, P., T. Mayer and F. Mayneris (2011), 'Spatial concentration and firm-level productivity in France', *Journal of Urban Economics*, **69**, 182–195.
Melo, P.C. and D.J. Graham (2009), 'Agglomeration economies and labour productivity: evidence from longitudinal worker data for GB's travel-to-work areas', SERC Discussion Paper 31.
Melo, P.C., D.J. Graham and R.B. Noland (2009), 'A meta-analysis of estimates of urban agglomeration economies', *Regional Science and Urban Economics*, **39**, 332–342.
Moran, P.A. (1950), 'Notes on continuous stochastic phenomena', *Biometrika*, **37**, 17–23.

Nocke, V. (2006), 'A gap for me: entrepreneurs and entry', *Journal of the European Economic Association*, **4** (5), 929–956.
Openshaw, S. and P. Taylor (1979), 'A million or so correlation coefficients: three experiments on the modifiable areal unit problem', in N. Wrigley (ed.), *Statistical Applications in the Spatial Sciences*, London: Pion, pp. 127–144.
Petrin, A., J. Levinsohn and B. Poi (2004), 'Production function estimation in stata using inputs to control for unobservables', *Stata Journal*, **4** (2), 113–123.
Puga, D. (2010) 'The magnitude and causes of agglomeration economies', *Journal of Regional Science*, **50** (1), 203–219.
Riley, S.J., S.D. DeGloria and R. Elliot (1999), 'A terrain ruggedness index that quantifies topographic heterogeneity', *Intermountain Journal of Science*, **5** (4), 23–27.
Rosenthal, S.S. and W.C. Strange (2004), 'Evidence on the nature and sources of agglomeration economies', in V. Henderson and J.F. Thisse (eds), *Handbook of Regional and Urban Economics*, Vol. 4, Amsterdam: North-Holland, pp. 2119–2171.
Stock, J.H. and J.H. Wright (2000), 'GMM with weak identification', *Econometrica*, **68**, 1055–1096.
Stock, J.H. and Yogo, M. (2002), 'Testing for weak instruments in linear IV regressions', NBER Technical Working Paper Series 284.
Syverson, C. (2011), 'What determines productivity?', *Journal of Economic Literature*, **49** (2), 326–365.
Venables, A.J. (2007), 'Evaluating urban transport improvements: cost–benefit analysis in the presence of agglomeration and income taxation', *Journal of Transport Economics and Policy*, **41** (2), 173–188.

APPENDIX

Table 10A.1 Sample: number of firms by year and total number of observations

Year	Number of firms
1997	11 328
1998	13 610
1999	29 318
2000	33 492
2001	40 942
2002	45 166
2003	46 366
2004	46 635
2005	48 703
Total	315 560

11. Accessibility: an underused analytical and empirical tool in spatial economics

Urban Gråsjö and Charlie Karlsson

11.1 INTRODUCTION

The so-called '1st law of geography' (Tobler, 1970) states that everything in space is related but that the relatedness between spatial units decreases with distance. This spatial dependence between spatial units should be perceived as a generic occurrence that is subject to distance-related friction phenomena. Spatial dependence implies for example that activities in one spatial unit have an effect on the activities in other regions but that the strength of this effect diminishes with distance. For example, spatial externalities that are mediated via the labour market depend on the interaction in the labour market – a market in which mobility is severely limited by the distance between spatial units. However, the spatial dependence between different spatial units also depends on the frequency of various types of interaction between these spatial units. That interaction decreases with distance is an axiomatic statement in regional science (cf. Beckmann, 2000). The accessibility approach offers an opportunity to develop measures that can account for the effect of distance-related frictions and thus how the strength of spatial dependencies diminishes with distance. Or in other words, accessibility measures approximate the potential for interaction among spatial units (Weibull, 1980). Accessibility measures represent spatial discounting procedures that relate to central concepts in spatial interaction theory.

The accessibility concept has a long history in both regional science and transport economics. According to Martellato et al. (1998), Hansen (1959) provided one of the first for the use of an 'accessibility theory' and defined accessibility as the potential of opportunities for interaction. Baradaran and Ramjerdi (2001) note that this way of defining accessibility is closely associated with gravity models based on the interaction of masses.

The purpose of this chapter is to show that accessibility is a useful analytical and empirical tool in spatial economics with an underestimated potential. We will not discuss alternative definitions and measures of accessibility and we will not try to review the general accessibility literature. There are already a substantial number of excellent reviews available (see for example Pirie, 1979; Handy and Niemayer, 1997; Reggiani, 1998). What we will do is to illustrate how accessibility measures can be used in a spatial context to explain patent output regional economic growth, new firm formation, the emergence of new export products, and so on. We will focus on empirical examples conducted in a Swedish context. The municipalities in Sweden are divided into local labour market regions[1] and this will affect how the accessibility measure is designed and used.

The chapter is structured as follows: section 11.2 introduces the accessibility concept and shows how it can be used to incorporate and explain spatial dependencies that may occur within regions and across regional borders. The section also demonstrates that an accessibility representation of explanatory variables depicts the network nature of spatial interaction, such that spatial dependence is actually modelled. Section 11.3 illustrates different settings where the accessibility concept can be or has been used in previous research. Section 11.4 concludes.

11.2 SPATIAL ENTITIES, TIME DISTANCES AND ACCESSIBILITY MEASURES[2]

The accessibility model presented in this chapter starts with the notion that a country can be divided into a number of labour market regions, each consisting of a number of municipalities between which the commuting intensity is high. In Sweden the delimitation of local labour markets is done in two steps:

1. Determination of local centres. Two conditions have to be fulfilled in order for a municipality to be a local centre:
   a. at least 80 per cent of the employed living in a municipality have to have their working place in the municipality;
   b. the number of commuters from a municipality to another municipality has to be below 7.5 per cent of the employed working force.
2. Determination of the remaining municipalities' belonging. The rest of the municipalities are connected directly or indirectly to the local centres that receive the largest number of commuters from these municipalities.

The number of local labour markets in Sweden has diminished over time, from 187 in 1970 to 79 in 2006. Consequently the average size of a local labour market has increased. Local labour markets can be found also in other countries. There are 15 countries within the European Union (EU) that use labour market areas. Usually these are built on the basis of municipalities. However, in Germany a level above municipalities is used and in Great Britain a level below the municipality level is used.

It is also possible to divide each municipality into a number of zones. From such a starting point we can imagine that it is meaningful to measure the accessibility between zones within a municipality, between municipalities within a local labour market region, and between a municipality in a given labour market region and all other municipalities in all other labour market regions in the country. In this manner it is possible to characterize the overall interaction patterns among spatial units, which naturally vary between different geographical scales and types of spatial units.

Accessibility can in this connection be thought of as a proximity measure to something desired (or something disliked for that matter). Thus, there are strong reasons to associate accessibility with preference or choice theory. Accessibility can be interpreted in several partly overlapping ways (Weibull, 1980): (1) nearness; (2) proximity; (3) the ease of spatial interaction; (4) potential opportunities of interaction; and (5) potential for contacts with activities (including supply and demand). Here the focus will be on interpretation (4) and how this interpretation can be related to preferences as specified in random choice theory.

Assume that an individual faces s choices, for example commuting links. We can then define an underlying latent variable U^*_{kl} to denote the level of indirect utility associated with the choice to commute from municipality k to municipality l. The observed variables U_{kl} are defined as:

$$U_{kl} = 1 \text{ if } U^*_{kl} = Max(U^*_{k1}, U^*_{k2}, \ldots, U^*_{ks}) \tag{11.1}$$

$U_{kl} = 0$ otherwise

Let us write $U^*_{kl} = V_{kl}(X_{kl}) + \varepsilon_{kl}$, where X_{kl} is a vector of attributes for choosing commuting link (k, l) and ε_{kl} is an extreme value distributed error term. Then it is possible to derive[3] the following probability that an individual in municipality k will choose the commuting link (k, l):[4]

$$P_{kl} = Prob(U_{kl} = 1|X) = exp\{V_{kl}\} / \sum_{s} exp\{V_{ks}\} \tag{11.2}$$

This formulation implies that the probability of choosing a specific link follows a Poisson distribution. In this case, the numerator in (11.2) represents the preference value of the labour market in municipality l and the denominator the sum of such values over all municipalities s. Thus, the probability of commuting on the link (k, l) is equal to the normalized preference value and P_{kl} can be interpreted as a ratio between the potential preference value of link (k, l) and the sum of preference values given by $\sum_s exp\{V_{ks}\}$.

Assume the following specification of the utility function:

$$V_{kl} = a_l - \gamma c_{kl} - \mu t_{kl} \tag{11.3}$$

where a_l represents an attractor factor in municipality l, c_{kl} denotes the commuting costs from k to l and t_{kl} is the time distance[5] between the municipalities. Let us now introduce two more assumptions: (1) $a_l = lnA_l$, where A_l represents the total number of jobs in l; and (2) $c_{kl} = \mu_c t_{kl}$, which implies that the commuting costs are proportional to the time distance on a link t_{kl}. With the use of these two assumptions the denominator in (11.2) can be expressed as:[6]

$$T_k^A = \sum_s exp\{-\lambda t_{ks}\} A_s \tag{11.4}$$

which is a standard measure of job accessibility in a municipality k, where the time sensitivity parameter $\lambda = (\gamma\mu_c + \mu)$. Based upon this formulation, it is now possible to define other accessibility measures, where the number of jobs A_s is substituted with other measures, such as the supply of household services, the supply of business services, the supply of labour, and so on in municipality s. Naturally, the opportunities are specific for each group of actors in the economy.

We are now in a position to ask to what extent interaction between zones within a municipality is different from interaction between municipalities in the same labour market region. Furthermore, is intra-regional interaction different from extra-regional interaction? The typical time distances for the three types of interaction in Sweden indicate that there may be a qualitative difference. For interaction between zones within municipalities the average time distance by car varies in the range 8–15 minutes. Inside a labour market region the average time distance by car has an interval of 20–50 minutes. Extra-regional time distances are, on average, longer than 60 minutes by car.

Given these travel time distances, it is natural to assume that the frequency of intra-municipality interactions between agents is much higher than the frequency of inter-municipality interactions, since mobility and interaction is time-consuming and also consumes other resources. Within the framework presented above this assumption can be taken care of by allowing the time-sensitivity parameter λ to be different for interactions inside a municipality than for interactions between municipalities. However, Johansson et al. (2002) have instead specified the attractiveness of the destination supply as different for intra- and extra-municipality interactions. They accomplish this as follows:

$$V_{kk} = ln\alpha_1 A_k - \lambda t_{kk} \text{ and}$$
$$V_{ks} = ln\alpha_2 A_s - \lambda t_{ks} \text{ for } s \neq k$$

where the first systemic preference indicator refers to intra-municipal interactions and the second to extra-municipal interactions. These indicators generate in a natural way a compound measure T_k^A of accessibility of municipality k:

$$T_k^A = \alpha_1 T_k^{AI} + \alpha_2 T_k^{AE}$$

where:

$$T_k^{AI} = A_k exp\{-\lambda t_{kk}\} \text{ and}$$
$$T_k^{AE} = \sum_{s \neq k} A_s exp\{-\lambda t_{rk}\}$$

represent intra-municipal and extra-municipal accessibility, respectively, and where $s \neq k$ is the set of municipalities except k.

Furthermore, it is possible to make a distinction between interactions that may occur between municipalities within the labour market region and accessibilities to all municipalities outside the region. If we also take into account the different time sensitivities three types of preference indicators can be identified:

$$V_{kk} = ln\alpha_1 A_k - \lambda_k t_{kk},$$
$$V_{ks}^R = ln\alpha_2 A_s - \lambda_R t_{ks} \text{ for } s \in R \text{ and}$$
$$V_{ks}^E = ln\alpha_3 A_s - \lambda_E t_{ks} \text{ for } s \in E$$

The compound measure T_k^A of accessibility of municipality k is given by:

$$T_k^A = \alpha_1 T_k^{AI} + \alpha_2 T_k^{AR} + \alpha_3 T_k^{AE}$$

where T_k^{AI} represents the intra-municipal accessibility of municipality k, T_k^{AR} represents the intra-regional accessibility of municipality k, that is, the accessibility to the other municipalities in the same labour market region

R, and T_k^{AE} represents the extra-regional accessibility of municipality k, that is, the accessibility to all municipalities outside the labour market region R.[7] Johansson et al. (2003) illustrate that the time sensitivities for the case of Sweden follow a non-linear form such that $\lambda_k < \lambda_E < \lambda_R$. Obviously, any accessibility for a municipality can be decomposed this way.

Potential statistical problems associated with dependence among observations in cross-sectional data are extensively treated in spatial econometrics literature (e.g. Anselin, 1988; Anselin and Florax, 1995; LeSage and Pace, 2009; Elhorst, 2010). A presence of any kind of spatial dependence can invalidate regression results. In the case of spatial error autocorrelation, ordinary least squares (OLS) parameter estimates are inefficient and in the presence of spatial lag dependence parameters become biased and inconsistent (Anselin, 1988). Moreover, a fundamental problem in applied spatial econometrics concerns the specification of the spatial interaction structure, that is, the structure of the spatial weight matrix (Florax and Rey, 1995). In the context of the present chapter, the inputs in other spatial units should optimally be spatially discounted in a way that reflects the distance sensitiveness of the effects (or externalities) involved. With respect to the spatial discounting procedure, this chapter advocates the use of accessibility as a measure of potential opportunities. Throughout the chapter, the spatial weight matrix is based on the concept of accessibility as a measure of potential opportunities.

Using the taxonomy by Anselin (2003), Andersson and Gråsjö (2009) investigate how the inclusion of spatially discounted variables (that is, accessibility variables) on the right-hand side (RHS) in empirical spatial models affects the extent of spatial autocorrelation. The basic proposition is that the inclusion of inputs external to the spatial observation in question as a separate variable reveals spatial dependence via the parameter estimate. One of the advantages of this method is that it allows for a direct interpretation. The authors also test to what extent significance of the estimated parameters of the spatially discounted explanatory variables can be interpreted as evidence of spatial dependence. Additionally, they advocate the use of the accessibility concept for spatial weights. Monte Carlo simulations show that the coefficient estimates of the accessibility variables are significantly different from zero. The rejection frequency of the three typical tests – Moran's I, LM-lag and LM-err – is significantly reduced, when accessibility variables are included in the model. The authors stress that when the coefficient estimates of the accessibility variables are statistically significant, it suggests that problems of spatial autocorrelation are significantly reduced. Significance of the accessibility variables can be interpreted as spatial dependence.

The accessibility approach is of great interest for policy-makers, since it makes it clear that there are two different ways to increase accessibility. Either the transport infrastructure and public transport can be improved to reduce travel times or potentials in different municipalities can be increased. However, the accessibility approach also makes it obvious that which transports links are improved and where the increased potentials are located is of great importance.

The accessibility variables can be calculated for different kinds of opportunities and used in empirical explanations of various spatial phenomena. The following section will illustrate how, for example, patent output, new firm formation, diversity of export products and economic growth in different spatial units have been modelled with the use of accessibility variables.

11.3 THE ACCESSIBILITY APPROACH IN DIFFERENT EMPIRICAL SETTINGS

The accessibility approach can be used in various situations. This section demonstrates its applicability and provides the reader with Swedish examples where the accessibility concept has been used in previous research concerning spatial economics.

11.3.1 Knowledge Production Functions

An accessibility approach to the analysis of knowledge spillovers has important implications for public policy. Because knowledge spillovers represent a positive externality and thus a disincentive for a firm to do research and development (R&D) and/or to produce at a socially optimal level, governments might use subsidies and other measures such as patent laws to encourage R&D and/or production. The framework presented in Karlsson and Manduchi (2001) offers a new perspective when discussing technology policy. It is obvious that technology policy must be discussed within this broader framework and not limited to issues regarding R&D and higher education. Also infrastructure policies involving local as well as intra- and interregional communication and transportation networks must be added to the agenda.

It is also obvious that simple solutions such as 'broadband Internet access for everyone' will not do the trick. There is a strong need to consider the complementarities between, on the one hand, communication and transportation networks, and on the other hand, between infrastructure investments and investments in R&D and higher education. One must in this connection also acknowledge that policies aiming at increasing knowledge spillovers to stimulate, for example, cluster formation may reduce the private incentives for doing R&D and hence demand either extended legal protection of inventions or larger public investments in or subsidies for R&D.

To model the influence of knowledge spillovers on knowledge production Griliches (1979) introduced the concept of a knowledge production function. The knowledge production function links the inputs in the innovation process to innovative outputs. According to Griliches, the most decisive innovative input is new economic knowledge and the greatest source of new economic knowledge is generally considered to be R&D. Jaffe (1989), Feldman (1994 a, 1994b) and Audretsch and Feldman (1996) modified the knowledge production function approach to a model specified for spatial and product dimensions.

The traditional knowledge production function approach tends to be used at an aggregate level and it does not consider the knowledge spillovers made possible by knowledge accessibility as defined here. Machlup (1980) defined knowledge production as any activity through which someone in a firm or an organization learns of something they had not known before, even if others knew about it. Knowledge production can involve both the creation of new knowledge and the search for new understanding from old knowledge. Knowledge production implicitly presumes the exchange of knowledge among persons. The formation of something new demands the amalgamation of different concepts and different pieces of knowledge. Such a creative feature of the process of knowledge exchange can be described as a form of dynamic synergy. Knowledge production activities hence demand a high degree of accessibility to other knowledgeable persons. We argue here that the output from R&D carried out within an industry or within specialized R&D institutions – that is, universities and similar institutions – is affected by knowledge spillovers. For the specialized R&D sector we assume that the important knowledge spillovers come on the one hand from within the sector, and on the other hand from other regions.

The link between proximity and innovation has been dwelt upon extensively in the literature. A regional economic milieu characterized by proximity between relevant actors is maintained to be suitable for establishing and maintaining successful regional innovation systems. Andersson and Karlsson (2004) propose that the relevant link to be studied is rather that between accessibility and innovation. The authors argue that although accessibility is a key factor in facilitating the processes important to innovation, the relationship is surprisingly unexploited.

Andersson and Ejermo (2002 [2003]) remark that knowledge production function (KPF) approaches to estimation of knowledge flows in regions have come under attack because they do not open the 'black box' of knowledge creation. It has been questioned whether spillovers really are the key determinants of knowledge diffusion rather than market mechanisms. Nonetheless, the authors claim that KPF approaches can be useful to get a rough picture of the aggregate magnitude of agglomeration effects pertaining to knowledge. Within a KPF framework, they study the relationship between the amount of R&D of firms and universities and the number of patent applications in Swedish functional regions. Interregional knowledge flows are weighted by the frictional effect of time distance. However, the analysis was not conducted within a proper spatial econometric framework and the authors therefore refrain from drawing any precise conclusions from the estimates.

In their 2004 paper Andersson and Ejermo (2004) attempt to explain knowledge production in Swedish functional regions as measured by the number of patent applications applying an accessibility approach. Recognizing that technological opportunity differs across sectors, a sectoral analysis is conducted. The KPF approach is applied in order to relate patent applications to a number of relevant knowledge sources. In addition to the R&D accessibility variable, the stock of patent applications is included as an explanatory variable in the analysis. The results show that the patent stock of a region contains much of the information needed in order to explain current patenting activity. This is interpreted as suggesting the strong effects of path dependence.

Andersson and Ejermo (2005) analyse the innovative performance of 130 Swedish corporations during 1993–1994 using an accessibility-based knowledge production function approach in line with that presented above. The number of patents per corporation is explained as a function of the accessibility to internal and external knowledge sources of each corporation. Their results show that there is a positive relationship between the innovativeness of a corporation and its accessibility to university researchers within regions where the corporation's research groups are located. The size of the corporation's R&D staff seems to be the most important internal factor. There is no indication in the results that intra-regional accessibility to other corporations' research is important for a corporation's innovativeness.

Ejermo and Gråsjö (2008) examine the effects of regional R&D on patenting for Sweden within an accessibility framework. They use two measures of patenting: number of patents granted per capita and a composite of quality adjusted patents which they regard as an innovation indicator, respectively. Two important conclusions emerge. Firstly, they find that the specification where innovations per capita is used as a dependent variable performs much better than with granted patents per capita for capturing relationships with regional R&D. Secondly, accessibility to interregional R&D does not affect innovation significantly, which suggests that effects are regionally bounded. This implies that studies of the R&D–innovation relationship are plagued by misspecification, since studies tend to show that R&D effects diffuse to other regions. This is also the case in their study: the interregional effects are an important factor for granted patents. In view of these results their recommendation is to use quality adjusted patents for regional innovation studies rather than patent grants.

The extent to which accessibility to R&D can explain patent production is further analysed by Gråsjö (2009). A knowledge production function is estimated both on an aggregate level and for different industrial sectors. The output of the knowledge production is the number of patent applications in Swedish municipalities from 1994 to 1999. The explanatory variables are expressed as accessibilities to university and company R&D at different spatial levels (local, intra-regional and interregional). A conclusion from the paper is that concentrated R&D investments in companies located in municipalities with a high patenting activity would benefit not only the municipalities themselves, but also the patent production in other municipalities in the same functional region.

The purpose of the paper by Gråsjö (2012) is to analyse the effects of national and international knowledge flows on innovative activity (patent applications in Swedish municipalities). The knowledge resources applied, R&D investments and high-valued imports, are expressed as accessibilities. The main results indicate that knowledge resources in a given municipality tend to have a positive effect on the innovative activity of other municipalities as long as they belong to the same functional region. This result holds for both R&D investments and high-valued import products.

11.3.2 Regional Productivity and Growth

Knowledge flows not only influence knowledge production, they also have a direct effect on the output of an individual industry in a region. Common output measures used in empirical studies that deal with regional productivity and growth are change in value added, gross regional product and gross pay. Other output measures like growth in population or employment can also be found.

Accessibility to knowledge and local service markets can be assumed to explain regional growth performance. The role of regional supply of services and educated labour with respect to regional development are stressed by many researchers. Karlsson and Pettersson (2005) make an empirical analysis using data for Swedish municipalities to analyse the relationship between regional productivity measures and accessibility to educated labour. They find that local externalities for increasing returns are very important in the Swedish economy. Their estimated models indicate that the elasticity for longer higher education and population density are around unity in the Swedish economy with respect to performance of regional gross domestic product per square kilometre.

Using an accessibility-based hierarchy of municipalities, Andersson and Klaesson (2006) relate growth in these municipalities to intra-municipal, intra-regional and interregional accessibility. They explore the growth in population, employment and commuting flows. The purpose of their study is to reveal systematic regularities in growth performance. Having established the overall pattern of change, they examine whether the information and communication technology (ICT) service sectors follow or deviate from this pattern. Their results show that there are strong similarities between the growth of individual ICT service sectors and the overall growth of the economy. Furthermore, the overall pattern suggests that municipalities with larger initial market accessibilities grow faster. This supports the presence of self-strengthening cumulative processes and implies that the size distribution of municipalities becomes more uneven over time.

The paper by Andersson and Noseleit (2009) also investigates Swedish employment growth. However, they extend previous analyses by examining the influence of regional start-ups in a sector on regional employment growth in the same sector and on other sectors. They find that knowledge-intensive start-ups seem to have larger effects on the regional economy. In particular, start-ups in high-end services have significant negative impacts on employment in other sectors but a positive long-run impact. This is consistent with the idea that start-ups are a vehicle for change in the composition of regional industry.

Several studies have been conducted on Swedish data to analyse the relationship between R&D investments and regional economic growth (Andersson et al., 2007; Andersson and Karlsson, 2007; Karlsson et al., 2008). Given the general assumption that R&D-generated knowledge contributes to economic growth, it is of great importance to understand how R&D contributes to economic growth in an economy where R&D is strongly concentrated into a limited number of regions. Strong evidence shows that knowledge transfers to a high extent depend upon face-to-face interaction, and the volume of knowledge flows depends upon the interaction possibilities at different spatial scales. It is meaningful to identify a number of such spatial scales based upon the character of the generalized spatial interaction costs. In particular, there are three spatial scales that are of special importance: (1) the local scale that allows several interactions a day; (2) the intra-regional or commuting scale that allows for daily interaction; (3) the interregional scale that allows only for a limited number of planned interactions per month or year. With the use of accessibility measures on these three scales it is possible to determine whether R&D generated knowledge has a local, intra-regional and/or interregional impact on economic growth. Two results stand out from the studies. Firstly, the knowledge accessibility in a given period has a statistically significant effect on the growth in subsequent periods. Secondly, the knowledge resources in a given municipality tend to have a positive effect on the growth of another municipality, conditional on the municipalities belonging to the same functional region. Knowledge flows thus transcend municipal borders, but they tend to be bounded within functional regions.

Andersson et al. (2008) focus on the role of human capital for regional productivity (gross pay per employee). They argue that a locality's position in a hierarchical spatial economic system is likely to alter the degree of impact human capital from the surrounding localities will have on its productivity level. The authors show that the relative importance of accessibility to external human capital for localities with a low position in a spatial hierarchy is significantly larger than for localities with a high position in the hierarchy.

It is well known that wages tend to be higher in larger regions. This can be explained by the fact that regions have different industrial compositions and that average regional productivity differs among regions. Using a decomposition method, similar to shift-share, Klaesson and Larsson (2009) separate regional wage differences into an industrial composition component and productivity component. According to the theory it is expected that productivity is higher in larger regions due to different kinds of economies of agglomeration. In addition, the diversity of sectors is more pronounced in larger regions compared to smaller regions. The authors use a market potential measure (accessibility to gross regional product) for regional size as a variable to explain regional differences in wages, productivity and industrial composition. Their results confirm that larger regions have higher wages, originating from higher productivity and more favourable industry composition.

Ejermo and Gråsjö (2011) explore the link between invention and innovation on the one hand, and the level of economic activity and economic growth in Swedish regions on the other, by using patents granted and the quality of patents as indicators of invention and innovation, respectively. Their results indicate that both types of measures are able to explain the level and the changing level of economic activity equally well. However, an important difference is that inventions have the strongest marginal effect in regions where economic activity is highest. Innovations have similar marginal effects across regions with different economic activity. The authors' interpretation is that quality-adjusted patents sort out 'bad' from 'good' patents in a manner which reflects economic importance.

11.3.3 The Relation between Company and University R&D

Rapid globalization in recent years has created a radically new competitive situation for the rich industrialized countries. Newly industrialized countries, and not least China, have become more and more successful in penetrating the markets in the rich industrialized countries with increasingly more advanced export products. This has generated a discussion in the rich industrialized countries on how to meet this increased international competition. In some countries demands for various protective measures have been raised while in others the discussion has mainly focused on how to develop a competitive strategy and have mainly focused on making their own products more sophisticated by increasing their knowledge content. This is by no means an easy task since direct product development is controlled to a high extent by multinational firms, which to an increasing degree are foreign-owned. Governments mainly have to rely on indirect measures, such as increasing the volume of higher education and public, mainly university, R&D. This raises the question of how responsive private industry is to these kinds of indirect measures.

Against this background, Andersson et al. (2009) present a study with the aim of analysing to what extent the location and the extent of higher education and university R&D, respectively, influence the location and the extent of industry R&D in Sweden, using an accessibility approach. They develop a model for the location of R&D from the perspective of a multinational enterprise and show that the location of industry R&D in Sweden can be partly explained by the intra-municipal accessibility to students in higher education, while the accessibility to university R&D turned out to be insignificant.

Karlsson and Andersson (2009) claim that while strong tendencies to greater globalization of R&D have been found, a strong spatial clustering of R&D and related innovative activities can also be observed. The standard explanation in the literature of the clustering of innovative activities is that such clusters offer external knowledge economies to innovative companies, since they are dependent upon knowledge flows, and that knowledge flows are spatially bounded. R&D is performed by two major players: industry and universities. It seems rather straightforward to assume that industrial R&D might be attracted to locate near research universities doing R&D in fields relevant to their industry. The question is whether it also works the other way around. Does industrial R&D function as an attractor for university R&D? It is possible to think of several reasons why university R&D may grow close to industry R&D. First of all, political decision-makers may decide to start or expand university R&D at locations where industry already is doing R&D work. Secondly, one can imagine that industry doing R&D in a region might use part of its R&D funds to finance university R&D. Thirdly, universities in regions with industrial R&D might find it easier to attract R&D funds from national and international sources due to cooperation with industry.

Obviously, not all types of university R&D attract industrial R&D. The above implies that there are behavioural relationships between industrial and university R&D. However, the literature contains few studies dealing with this problem. Most studies have concentrated on the one-directional effect from university R&D to industrial R&D, and the outputs of industrial R&D in most cases measured in terms of the number of patents, while neglecting the possible mutual interaction. However, if there is a mutual interaction between university and industry R&D and if knowledge externalities are involved, then it is possible, as Karlsson and Andersson (2009) have done, to develop a dynamic explanation for the clustering of innovative activities based on positive feedback loops. This implies strong tendencies to path-dependency and that policy initiatives to transform non-innovative regions into innovative regions would have little chance of success. Karlsson and Andersson (2009) show that the location of industrial R&D is sensitive to the accessibility of university R&D, and vice versa.

11.3.4 Exports

The relation between export competitiveness and knowledge at both the national and the firm level is explored in several empirical studies (Fagerberg, 1988; Greenhalgh et al., 1994; Wakelin, 1998; Basile, 2001). The general conclusions from these studies are that innovation, measured using proxies for inputs (for example, R&D expenditure) or for outputs (for example, number of patents) is an important factor in explaining export performance. However, what is lacking in the studies at national and firm level is the role of geographical proximity in facilitating the transmission and the absorption of technological and scientific knowledge.

Johansson and Karlsson (2007) examine the influence of accessibility to R&D on regional diversity in Swedish exports. They argue that the effects of R&D on regional export performance are reflected by the size of the export base rather than by the export volumes. The empirical analysis focuses on three different indicators of export diversity: the number of exported goods, the number of exporting firms and the number of export destinations. The results suggest that the three indicators of regional export diversity are positively affected by the intra-regional accessibility to company R&D in commodity groups where production is relatively R&D intensive. Interregional accessibility to company R&D has significant positive impacts on the number of export goods and the number of export destinations also in less R&D-intensive industries. In the case of university R&D, the empirical results are weaker, in particular in the case of intra-regional accessibility. Yet the interregional accessibility to university R&D has a significant positive impact on the number of export goods and the number of export destinations in the majority of commodity groups.

The extent to which accessibility to R&D and human capital can explain regional exports is also analysed by Gråsjö (2008). The author performs a comparison between a volume measure (total export value) and a diversity measure (number of high-value export products) in Swedish municipalities. The results indicate that accessibility to human capital has the greatest positive effects. The value of exported products is mainly affected by local accessibility to human capital (and company R&D). The intra- and interregional accessibilities play a more important role when the output being considered is the number of high-value export products in Swedish municipalities.

Bjerke and Karlsson (2009), on the other hand, focus on the role that metropolitan regions play in the renewal of the export base in the non-metropolitan regions in a small country. In smaller countries, the non-metropolitan regions are to a large extent linked together with the metropolitan regions through various networks. The national infrastructure and transport networks are often organized with the metropolitan region as the central hub. This creates a number of dependencies between the metropolitan region and the non-metropolitan regions in a small country. The analytical part of their paper can be divided into three main sections: (1) the role of the Stockholm metropolitan region for the renewal of the export base in the rest of Sweden between 1997 and 2003; (2) which non-metropolitan regions gain renewal of their export base; and (3) what factors can explain the spatial distribution of these gains. The results of the paper show that distance has little to do with the potential success of export products diffused from Stockholm. Instead, regional characteristics such as a large manufacturing sector, educational level, size of public and/or agricultural sector, and high intra-regional accessibility to producer services have a larger influential potential.

11.3.5 New Firm Formation

Several scholars have included accessibility-based measures in their analysis of factors determining new firm formation. Andersson and Hellerstedt (2009) study start-ups in knowledge-intensive business services (KIBS) across regions in Sweden. Their empirical analysis takes into account both supply- and demand-side factors. Supply-side variables reflect knowledge and information upon which a new firm can be established. Demand-side variables refer to market potential proxied by accessibility to regional gross pay. Controlling for the stock of potential entrepreneurs and the stock of KIBS firms, they show that both supply- and demand-side factors influence KIBS start-up activity. Their results display that the presence of knowledge resources and accessibility to a large market are beneficial conditions for KIBS start-ups.

Karlsson and Backman (2011) investigate the impact of human capital accessibility on new firm formation. The empirical analysis in the study is based on data on new firm formation at the municipality level in Sweden and accessibility to human capital, where 'carriers of human capital' is defined as those individuals with at least three years of university education. The results indicate that intra-municipal accessibility to human capital has a positive impact on new firm formation in municipalities.

Accessibility-based measures to explain new firm formation are also used by Grek et al. (2011). The authors' purpose is to explain the variations in entrepreneurship between regions of various sizes and to test the theoretical arguments on why large regions generally should generate more entrepreneurship. The results show that the market potential as measured by local and external accessibility to gross regional product (GRP) has a strong significant impact both on entry of new firms and on firm exit. For the primary sector and the manufacturing sector this impact is negative, while for the ordinary service sector and the advanced service sector it is positive. A high employment rate has a strong negative impact on firm entry in all sectors. This is in line with what one could expect as there are weaker incentives for individuals starting their own businesses in periods of low unemployment. Furthermore, the presence of many small firms in different sectors has a strong positive significant impact on new firm formation. Andersson and Koster (2011) also make use of accessibility to GRP as a measure of regional market potential. The paper analyses the persistence of start-up rates across Swedish regions. The authors find that start-up rates of a decade earlier are able to explain over 40 per cent of the variation in current start-up rates across regions.

Karlsson and Nyström (2011) investigate the role of accessibility to university and company R&D for new firm formation. Company R&D

is assumed to contain a higher share of R&D directed towards generating technological knowledge. Hence, the accessibility to such R&D is expected to have a stronger influence on new firm formation than the accessibility to university R&D, and this is also what the empirical results of the paper indicate. The authors also find that close knowledge interactions are more important for new firm formation than long-distance knowledge interactions. Accessibility to interregional company R&D even has a negative impact on new firm formation.

11.3.6 Regional Interaction and Diversity

A paper by Andersson and Klaesson (2009) analyses how a region's relative market accessibility in a system (or hierarchy) of municipalities affects the extent of diversity. In the theoretical part of the paper a model of municipal diversity in retail and durables is introduced. Using this model as a point of reference, the authors explore the relationship between market size and diversity in Swedish regions. Three types of market sizes are considered: intra-municipal, intra-regional and extra-regional. They show that the relationships between diversity and the three types of market sizes differ between different types of municipalities in the hierarchy, implying that such a classification is warranted. One particular finding that corresponds to the agglomeration shadow effects usually discussed in NEG theories is that large municipalities gain from proximity to surrounding municipalities while small municipalities do not.

11.3.7 Location Dynamics of Firms

Andersson (2006) investigates the tendencies of co-location between producer services and manufacturing across Swedish functional regions using an accessibility-based approach. The employment in these industries is modelled simultaneously, that is, the location of producer services is a function of the accessibility of manufacturing and vice versa. The assumption motivating the simultaneous approach is that manufacturing firms benefit from short-distance supply of producer services, and service suppliers benefit from accessibility to customers among the manufacturing firms. The empirical results of the paper suggest that the location of manufacturing employment can be explained by its accessibility to producer services. However, accessibility to manufacturing is not a statistically significant explanatory factor for the location of producer services.

Johansson and Klaesson (2011) consider the location dynamics of two categories of firms: contact-intensive producer-service suppliers and other firms. The authors argue that firms have random choice preferences and react in a non-linear way to time distances in their contact efforts. Hence, firms make their location decisions in response to local, intra-regional and interregional accessibility to market demand. The econometric analysis in the paper takes into account time distances between zones in urban areas as well as between urban areas in the same agglomeration and between urban areas in different agglomerations. This information is used in an econometric model that depicts for each urban region how the number of jobs in different sectors changes in response to the access to customers' purchasing power in the entire set of urban regions. The empirical results of the paper suggest that firms' location choices depend on local and intraregional accessibility to market demand. Interregional accessibility is also of importance, but only for producer-service suppliers and not for other firms.

11.3.8 Labour Mobility

Andersson and Thulin (2011) focus on interfirm labour mobility. They study to what extent spatial employment density can explain interfirm job switching. The empirical results of the study show that employment density has a positive impact on the probability of job switching and that interfirm labour mobility varies substantially across regions. Moreover, the likelihood that such switching is intra-regional is significantly higher if the employees operate in denser regions. The authors conclude that higher rates of interfirm labour mobility seem to be a probable mechanism behind the empirically verified productivity advantage of dense regions.

11.3.9 Summary of the Empirical Studies

As illustrated, the accessibility concept can be used in numerous empirical settings. The main research questions dealt with in this section are to what extent:

1. regional patent production is explained by accessibility to knowledge resources (mainly R&D, but also diversity of import products);
2. regional productivity and growth (employment, gross pay, value added, and so on) is affected by accessibility to knowledge resources (R&D, educated labour, patents) and market size;
3. regional diversity in exports is influenced by accessibility to R&D, educated labour and producer services;
4. regional start-up rates are dependent on accessibility to market size (gross regional product and gross pay) and R&D;

5. location decisions made by firms are explained by accessibility to market size (gross pay, producer services and manufacturing).

Table 11.1 presents, in brief, the empirical studies included in this chapter.

In light of generalization issues and related policy perspectives, it would be interesting for future research to apply the accessibility approach, applied in the various Swedish case studies, to different spatial contexts and landscapes. It is a general approach, because whenever the theory suggests that inputs from outside a location are assumed to have an impact on its output, but that these interlocational effects diminish with distance, the accessibility measure is a potentially useful tool. The accessibility approach is also of great interest for public policy. In order to achieve increased accessibility policy-makers can either enhance the potential of locations or improve the transport infrastructure between and within locations. In addition, since the approach takes into account different spatial levels, infrastructure policies involving local, intra-regional and interregional transportation networks can be compared and valued.

11.4 CONCLUSIONS

The purpose of this chapter was to show that the accessibility approach is a very useful analytical and empirical tool in spatial economics. We have illustrated how accessibility measures can be used in a spatial context to explain numerous economic phenomena, such as patent output, regional economic growth, new firm formation, the diversity of exports, and so on.

The chapter promotes the use of the accessibility concept for several reasons:

1. It is related to spatial interaction theory and can be motivated theoretically by adhering to the preference structure in random choice theory.[8]
2. It incorporates 'global' spillovers and does not only account for the impact from neighbours or locations within a certain distance band.
3. The separation into local, intra-regional and interregional accessibilities captures potential productive dependencies between locations and makes the inferential aspects more clear.
4. Distance is often measured by the physical distance, but a more appropriate and realistic measure in economic modelling is the time it takes to travel between different locations. The accessibility measure in this chapter is constructed with the use of commuting time distance.

Table 11.1 Empirical studies using the accessibility approach

	Dependent variable	Accessibility variables (independent)	Unit of analysis
Knowledge production			
Andersson and Ejermo (2004)	Patents	Company and university R&D	Functional regions
Andersson and Ejermo (2005)	Patents per corporation	Company and university R&D	Functional regions
Ejermo and Gråsjö (2008)	Patents per cap, Quality adjusted patents per cap	Company and university R&D	Functional regions
Gråsjö (2009)	Patents	Company and university R&D	Municipalities
Gråsjö (2012)	Patents	R&D, High valued imports	Municipalities
Productivity and growth			
Karlsson and Pettersson (2005)	Gross regional product per km^2	Population, high-educated labour	Municipalities
Andersson and Klaesson (2006)	Growth in population, employment, commuting	Market size (gross pay)	Municipalities
Andersson et al. (2007)	Growth in value added per employee	Company and university R&D	Municipalities
Andersson and Karlsson (2007)	Growth in value added per employee	Knowledge resources	Municipalities
Andersson et al. (2008)	Gross pay per employee	University-educated labour	Municipalities
Karlsson et al. (2008)	Growth in value added per employee	Company and university R&D	Municipalities
Andersson and Noseleit (2009)	Employment growth	Firm start-ups	Functional regions
Klaesson and Larsson (2009)	Gross pay per employee, productivity and industrial composition index	Market potential (gross regional product)	Municipalities
Ejermo and Gråsjö (2011)	Change in gross regional product and gross pay	Patents, quality adjusted patents	Functional regions

Table 11.1 (continued)

	Dependent variable	Accessibility variables (independent)	Unit of analysis
Company and university R&D			
Andersson et al. (2009)	Company R&D	University R&D, higher education	Municipalities
Karlsson and Andersson (2009)	Company R&D, University R&D	University R&D, Company R&D	Municipalities
Exports			
Johansson and Karlsson (2007)	Exported goods, exporting firms, export destinations	Company and university R&D	Functional regions
Gråsjö (2008)	Total export value, high-value exports	R&D, human capital	Municipalities
Bjerke and Karlsson (2009)	Export value	Producer services	Functional regions
New firm formation			
Andersson and Hellerstedt (2009)	KIBS start-ups	Market potential (gross pay)	Municipalities
Karlsson and Backman (2011)	Start-ups	Human capital	Municipalities
Grek et al. (2011)	Entry and exit of firms	Gross regional product	Municipalities
Andersson and Koster (2011)	Start-ups	Gross regional product	Municipalities
Karlsson and Nyström (2011)	Start-ups	Company and university R&D	Municipalities
Interaction and diversity			
Andersson and Klaesson (2009)	Entropy of establishments	Population	Municipalities
Location dynamics of firms			
Andersson (2006)	Employment, manufacturing and producer services	Employment, producer services Employment manufacturing	Functional regions
Johansson and Klaesson (2011)	Change in number of jobs	Market demand (gross pay)	Functional regions
Labour mobility			
Andersson and Thulin (2011)	Labour mobility	Employment	Functional regions

5. Econometric problems with biased parameter estimates are reduced even if the underlying spatial structure is expressed by spatially lagged dependent variables. In addition, the parameter estimates are much more efficient when the accessibility variables are included in the model.
6. The accessibility approach is of great interest for policy-makers, since it makes clear that improved accessibility can be achieved in two ways. Either policy-makers can improve the transport infrastructure and public transport to reduce travel times, or they can increase the potentials in different municipalities. However, the accessibility approach also makes it obvious that it is of great importance which transport links are improved and where the increased potentials are located.

Being an underused analytical and empirical tool in spatial economics, we welcome more research that uses the accessibility approach in the future.

NOTES

1. The concept of a local labour market region is closely associated with the concept of a functional urban region (cf. Cheshire and Gordon, 1998).
2. This section builds upon Johansson et al. (2002) and Andersson and Gråsjö (2009).
3. See for example Maddala (1983) or Train (1986).
4. We use commuting between municipalities as an example but we could have used any type of interaction as our example. The conclusions are general.
5. Researchers often measure distance by the geographical distance, but a better way to measure it is to use the time it takes to travel between different locations (Beckmann, 2000). Time distance is, for example, crucial for the frequency of interregional business trips in Sweden (Hugosson, 2001; Johansson et al., 2002).
6. The negative exponential function emerges directly from an entropy maximizing framework with origin, destination and cost constraints (see Smith, 1978; Wilson, 2000).
7. The accessibility measures used here satisfy criteria of consistency and meaningfulness (Weibull, 1976) and have a clear coupling to spatial interaction theory.
8. Fingleton (2003) remarks that the spatial weight matrix applied in many empirical studies is not underwritten by a strong theory and that the assumptions behind the chosen weight matrix are often not tested.

REFERENCES

Andersson, M. (2006), 'Co-location of manufacturing and producer services – a simultaneous equation approach', in C. Karlsson, B. Johansson and R.R. Stough (eds), *Entrepreneurship and Dynamics in the Knowledge Economy*, New York: Routledge, pp. 94–124.

Andersson, M. and O. Ejermo (2002 [2003]), 'Knowledge production in Swedish functional regions 1993–1999', KITeS Working Paper 139, KITeS, Centre for Knowledge, Internationalization and Technology Studies, Università Bocconi, Milano, Italy; revised February 2003.

Andersson, M. and O. Ejermo (2004), 'Sectoral knowledge production in Sweden 1993–1999', in C. Karlsson, P. Flensburg and S.-Å. Hörte (eds), *Knowledge Spillovers and Knowledge Management*, Cheltenham, UK and Northampton, MA, USA: Edward Elgar, pp. 143–170.

Andersson, M. and O. Ejermo (2005), 'How does accessibility to knowledge sources affect the innovativeness of corporations? Evidence from Sweden', *Annals of Regional Science*, **39** (4), 741–765.

Andersson, M. and U. Gråsjö (2009), 'Spatial dependence and the representation of space in empirical models', *Annals of Regional Science*, **43**, 159–180.

Andersson, M., U. Gråsjö and C. Karlsson (2007), 'Regional growth and accessibility to knowledge resources: a study of Swedish municipalities', *ICFAI Journal of Knowledge Management*, July, 7–26.

Andersson, M., U. Gråsjö and C. Karlsson (2008), 'Human capital and productivity in spatial economic systems', *Annales d'Economie et de Statistique*, **87–88**, 125–143.

Andersson, M., U. Gråsjö and C. Karlsson (2009), 'The role of higher education and university R&D for industrial R&D location', in A. Varga (ed.), *Universities, Knowledge Transfer and Regional Development. Geography, Entrepreneurship and Policy*, Cheltenham, UK and Northampton, MA, USA: Edward Elgar, pp. 85–108.

Andersson, M. and K. Hellerstedt (2009), 'Location attributes and start-ups in knowledge-intensive business services', *Industry and Innovation*, **16** (1), 103–121.

Andersson, M. and C. Karlsson (2004), 'The role of accessibility for the performance of regional innovation systems', in C. Karlsson, P. Flensburg and S.-Å. Hörte (eds), *Knowledge Spillovers and Knowledge Management*, Cheltenham, UK and Northampton, MA, USA: Edward Elgar, pp. 311–346.

Andersson, M. and C. Karlsson (2007), 'Knowledge in regional economic growth – the role of knowledge accessibility', *Industry and Innovation*, **14**, 129–149.

Andersson, M. and J. Klaesson (2006), 'Growth dynamics in a municipal marker accessibility hierarchy', in B. Johansson, C. Karlsson and R.R. Stough (eds), *The Emerging Digital Economy: Entrepreneurship, Clusters and Policy*, Berlin: Springer, pp. 187–212.

Andersson, M. and J. Klaesson (2009), 'Regional interaction and economic diversity – exploring the role of geographically overlapping markets for a municipality's diversity in retail and durables', in C. Karlsson, B. Johansson and R.R. Stough (eds), *Innovation, Agglomeration and Regional Competition*, Cheltenham, UK and Northampton, MA, USA: Edward Elgar, pp. 19–37.

Andersson, M. and S. Koster (2011), 'Sources of persistence in regional start-up rates – evidence from Sweden', *Journal of Economic Geography*, **11**, 179–201.

Andersson, M. and F. Noseleit (2009), 'Start-ups and employment dynamics within and across sectors', *Small Business Economics*, **36** (4), 461–483.

Andersson, M. and P. Thulin (2011), 'Labour mobility and spatial density', CESIS Working Paper 248, CESIS, KTH, Stockholm and JIBS, Jönköping.

Anselin, L. (1988), *Spatial Econometrics: Methods and Models*, Boston, MA: Kluwer Academic Publishers.

Anselin, L. (2003), 'Spatial externalities, spatial multipliers and spatial econometrics', *International Regional Science Review*, **26**, 153–166.

Anselin, L. and R. Florax (eds) (1995), *New Directions in Spatial Econometrics*, Berlin: Springer Verlag.
Audretsch, D.P. and M.P. Feldman (1996), 'R&D spillovers and the geography of innovation and production', *American Economic Review*, **86**, 630–640.
Baradaran, S. and F. Ramjerdi (2001), 'Performance of accessibility measures in Europe', *Journal of Transport Statistics*, **3**, 31–48.
Basile, R. (2001), 'Export behaviour of Italian manufacturing firms over the nineties: the role of innovation', *Research Policy*, **30**, 1185–1201.
Beckmann, M. (2000), 'Interurban knowledge networks', in D.F. Batten (ed.), *Learning, Innovation and the Urban Evolution*, London: Kluwer Academic Publishers, 127–135.
Bjerke, L. and C. Karlsson (2009), 'Metropolitan regions and product innovation', CESIS Working Paper 166, CESIS, KTH, Stockholm and JIBS, Jönköping.
Cheshire, P.C. and I.R. Gordon (1998), 'Territorial competition: some lessons for policy', *Annals of Regional Science*, **32**, 321–346.
Ejermo, O. and U. Gråsjö (2008), 'The effects of R&D on regional invention and innovation', Circle Working Paper 2008/03, University of Lund.
Ejermo, O. and U. Gråsjö (2011), 'Invention, innovation and regional growth in Swedish regions', in B. Johansson, C. Karlsson and R. Stough (eds), *Innovation, Technology and Knowledge: Their Role in Economic Development*, London: Routledge, pp. 187–208.
Elhorst, J.P. (2010), 'Applied spatial econometrics: raising the bar', *Spatial Economic Analysis*, **5** (1), 9–28.
Fagerberg, J. (1988), 'International competitiveness', *Economic Journal*, **98**, 355–374.
Feldman, M.P. (1994a), *The Geography of Innovation*, Boston, MA: Kluwer Academic Publishers.
Feldman, M.P. (1994b), 'Knowledge complementarity and innovation', *Small Business Economics*, **6**, 363–372.
Fingleton, B. (2003), 'Externalities, economic geography and spatial econometrics: conceptual and modelling developments', *International Regional Science Review*, **26**, 197–207.
Florax, R. and S. Rey (1995), 'The impacts of misspecified spatial interaction in linear regression models', in L. Anselin and R. Florax (eds), *New Directions in Spatial Econometrics*, Berlin: Springer Verlag, pp. 111–113.
Greenhalgh, C., P. Taylor and R. Wilson (1994), 'Innovation and export volumes and prices, a disaggregated study', *Oxford Economic Papers*, **46**, 102–134.
Grek, J., C. Karlsson and J. Klaesson (2011), 'Determinants of entry and exit: the significance of demand and supply conditions at the regional level', in K. Kourtit, P. Nijkamp and R.R. Stough (eds), *Drivers of Innovation, Entrepreneurship and Regional Dynamics*, Berlin: Springer, pp. 121–141.
Griliches, Z. (1979), 'Issues in assessing the contribution of R&D to productivity growth', *Bell Journal of Economics*, **10**, 92–116.
Gråsjö, U. (2008), 'University educated labour, R&D and regional export performance', *International Regional Science Review*, **31** (3), 211–256.
Gråsjö, U. (2009), 'Accessibility to R&D and patent production', in C. Karlsson, B. Johansson and R.R. Stough (eds), *Innovation, Agglomeration and Regional Competition*, Cheltenham, UK and Northampton, MA, USA: Edward Elgar, pp. 231–260.
Gråsjö, U. (2012), 'Imports, R&D and local patent production', in C. Karlsson,

B. Johansson and R. Stough (eds), *The Regional Economics of Knowledge and Talent. Local Advantage in a Global Context*, Cheltenham, UK and Northampton, MA, USA: Edward Elgar Publishing, pp. 343–366.
Handy, S.L. and D.A. Niemayer (1997), 'Measuring accessibility: an exploration of issues and alternatives', *Environment and Planning A*, **29**, 1175–1194.
Hansen, W.G. (1959), 'How accessibility shapes land-use', *Journal of the American Institute of Planners*, **25**, 73–76.
Hugosson, P. (2001), *Interregional Business Travel and the Economics of Business Interaction*, Jönköping: Jönköping International Business School Dissertation Series No. 009.
Jaffe, A.B. (1989), 'Real effects of academic research', *American Economic Review*, **79**, 957–970.
Johansson, B. and J. Klaesson (2011), 'Agglomeration dynamics of business services', *Annals of Regional Science*, **47** (2), 373–391.
Johansson, B., J. Klaesson and M. Olsson (2002), 'Time distances and labour market integration', *Papers in Regional Science*, **81**, 305–327.
Johansson, B., J. Klaesson and M. Olsson (2003), 'Commuters' non-linear response to time distances', *Journal of Geographical Systems*, **5**, 315–329.
Johansson, S. and C. Karlsson (2007), 'R&D accessibility and regional export diversity', *Annals of Regional Science*, **41**, 501–523.
Karlsson, C. and M. Andersson (2009), 'The location of industry R&D and the location of university R&D: how are they located?', in C. Karlsson, A.E. Andersson, P.C. Cheshire and R.R. Stough (eds), *New Directions in Regional Economic Development*, Berlin: Springer, pp. 267–290.
Karlsson, C., M. Andersson and U. Gråsjö (2008), 'University and industry R&D accessibility and regional growth', *Italian Journal of Regional Science*, **7**, 97–117.
Karlsson, C. and B. Backman (2011), 'Accessibility to human capital and new firm formation', *International Journal of Foresight and Innovation Policy*, **7**, 7–22.
Karlsson, C. and A. Manduchi (2001), 'Knowledge spillovers in a spatial context – a critical review and assessment', in M.M. Fischer and J. Fröhlich (eds), *Knowledge, Complexity and Innovation Systems*, Berlin: Springer-Verlag, pp. 101–123.
Karlsson, C. and K. Nyström (2011), 'Knowledge accessibility and new firm formation', in S. Desai, P. Nijkamp and R.R. Stough (eds), *New Directions in Regional Economic Development. The Role of Entrepreneurship Theory and Methods, Practice and Policy*, Cheltenham, UK and Northampton, MA, USA: Edward Elgar, pp. 174–197.
Karlsson, C. and L. Pettersson (2005), 'Regional productivity and accessibility to knowledge and dense markets', CESIS Working Paper 32, CESIS, KTH, Stockholm and JIBS, Jönköping.
Klaesson, J. and H. Larsson (2009), 'Wages, productivity and industry composition – agglomeration economies in Swedish regions', CESIS Working Paper 203, CESIS, KTH, Stockholm and JIBS, Jönköping.
LeSage, J.P. and R.K. Pace (2009), *Introduction to Spatial Econometrics*, Boca Raton, FL: Taylor & Francis.
Machlup, F. (1980), *Knowledge and Knowledge Production*, Princeton, NJ: Princeton University Press.
Maddala, G.S. (1983), *Limited Dependent and Quantitative Variables in Econometrics*, Cambridge: Cambridge University Press.
Martellato, D., P. Nijkamp and A. Reggiani (1998), 'Measurement and measures

of network accessibility: economic perspectives', in K. Button, P. Nijkamp and H. Priemus (eds), *Transport Networks in Europe: Concepts, Analysis and Policies*, Cheltenham, UK and Northampton, MA, USA: Edward Elgar, pp. 161–180.

Pirie, G.H. (1979), 'Measuring accessibility: a review and proposal', *Environment and Planning A*, **11**, 299–312.

Reggiani, A. (ed.) (1998), *Accessibility, Trade and Location Behaviour*, Aldershot: Ashgate.

Smith, T.E. (1978), 'A cost-efficiency principle of spatial interaction behaviour', *Regional Science and Urban Economics*, **8**, 313–337.

Tobler, W.R. (1970), 'A computer movie simulating urban growth in the Detroit region', *Economic Geography*, **46**, 234–240.

Train, K. (1986), *Qualitative Choice Analysis*, Boston, MA: MIT University Press.

Wakelin, K. (1998), 'Innovation and export behaviour at the firm level', *Research Policy*, **26** (7–8), 829–841.

Weibull, J. (1976), 'An axiomatic approach to the measurement of accessibility', *Regional Science and Urban Economics*, **6**, 357–379.

Weibull, J.W. (1980), 'On the numerical measurement of accessibility', *Environment and Planning A*, **12**, 53–67.

Wilson, A.G. (2000), *Complex Spatial Systems – The Modelling Foundations of Urban and Regional Analysis*, London: Prentice Hall.

Index